INNATE

INNATE

How the Wiring of Our Brains Shapes Who We Are

KEVIN J. MITCHELL

PRINCETON UNIVERSITY PRESS

PRINCETON AND OXFORD

Published by Princeton University Press
41 William Street, Princeton, New Jersey 08540
6 Oxford Street, Woodstock, Oxfordshire OX20 1TR

press.princeton.edu

Library of Congress Control Number: 2018937061
ISBN 978-0-691-17388-7

British Library Cataloging-in-Publication Data is available

Editorial: Alison Kalett and Lauren Bucca
Production Editorial: Jenny Wolkowicki
Text and Jacket Design: Lorraine Doneker
Production: Jacqueline Poirier
Publicity: Sara Henning-Stout
Copyeditor: Maia Vaswani

This book has been composed in Minion Pro with Din for display

Printed on acid-free paper. ∞

Printed in the United States of America

10 9 8 7 6 5 4 3 2 1

For my parents,
with my undying thanks for their nature and their nurture.

CONTENTS

ACKNOWLEDGMENTS

I am very grateful to numerous friends and colleagues for thoughtful discussions over many years, which have informed the ideas presented in this book, including Cori Bargmann, Alain Chedotal, Aiden Corvin, Kevin Devine, John Foxe, Michael Gill, Corey Goodman, Josh Gordon, Josh Huang, Andrew Jackson, Annette Karmiloff-Smith, Mary-Claire King, David Ledbetter, Gary Marcus, Oscar Marin, Aoife McLysaght, Partha Mitra, Bita Moghaddam, Scott Myers, Fiona Newell, Shane O'Mara, Mani Ramaswami, Ian Roberston, John Rubenstein, Carla Shatz, Bill Skarnes, Marc Tessier-Lavigne, and many others, especially all the past and current members of my lab. I should include in this list a huge number of people who have been kind enough to engage in scientific or philosophical discussions on the *Wiring the Brain* blog or on Twitter, which have been invaluable in enriching and sharpening my thinking. I am especially grateful to Abeba Birhane, Sarah-Jayne Blakemore, Dan Bradley, David Delany, Adam Kepecs, Tim Lewens, David McConnell, Lynn Mitchell, Tara Mitchell, Thomas Mitchell, Svetlana Molchanova, Stuart Ritchie, Richard Roche, Siobhan Roche, and Adam Rutherford for very helpful feedback on chapters of the book. I would also like to sincerely thank the three anonymous reviewers for their support and valuable suggestions. And I am extremely grateful for all the support from Princeton University Press, especially to my editor Alison Kalett, for her encouragement, insight, and expert advice throughout the process of writing this book.

INNATE

ON HUMAN NATURE

How would you describe yourself? If you had to list some personality traits, say for a dating website or a job application, what words would you use? Do you consider yourself shy or outgoing? Are you cautious or reckless? Anxious or carefree? Are you creative, artistic, adventurous, stubborn, impulsive, sensitive, brave, mischievous, kind, imaginative, selfish, irresponsible, conscientious? People clearly differ in such traits and in many other aspects of their psychology—such as intelligence and sexual preference, for example. All of these things feed into making us *who we are*.

The question is, how do we get that way? This has been a subject of endless debate for literally thousands of years, with various prominent thinkers, from Aristotle and Plato to Pinker and Chomsky, lining up to argue for either innate differences between people or for everyone starting out with a blank slate and our psychology being shaped by experience alone. In the past century, the tradition of Freudian psychology popularized the idea that our psychological dispositions could be traced to formative childhood experiences. In many areas of modern academic sociology and psychology this belief is still widespread, though it has been extended to include cultural and environmental factors more broadly as important determinants of our characters.

But these fields have been fighting a rearguard action in recent years, against an onslaught from genetics and neuroscience, which have provided strong evidence that such traits have at least some basis in our innate biology. To some, this is a controversial position, perhaps even a morally offensive one. But really it fits with our common experience that, at some level, people just are the way they are—that they're just made that way. Certainly, any parent with more than one child will know that they start out different from each other, in many important ways that are unrelated to parenting.

This notion of innate traits is often equated with the influences of genes—indeed, "innate" and "genetic" are often used interchangeably. This idea is captured in common phrases such as "the apple doesn't fall far from the tree," or "he didn't lick it off the stones." These sayings reflect the widespread belief that many of our psychological traits are not determined solely by our upbringing but really are, to some extent at least, "in our DNA."

How that could be is the subject of this book. How could our individual natures be encoded in our genomes? What is the nature of that information and how is it expressed? That is, in a sense, just a different version of this question: How is human nature, generally, encoded in the human genome? If there is a program for making a human being with typical human nature, then our individual natures may simply be variations on that theme. In the same way, the human genome contains a program for making a being about so tall, but *individual humans* are taller or shorter than that due to variation in the programs encoded in their respective genomes. We will see that the existence of such variation is not only plausible—it is inevitable.

BEING HUMAN

If we think about human nature generally, then we should ask, first, whether it even exists. Are there really typical characteristics that are inherent in each of us that make humans different from other animals? This question has exercised philosophers for millennia and continues to today, partly because it can be framed in many different ways. By human nature, do we mean expressed behaviors that are unique to humans and not seen in other animals? Do we mean ones that are completely universal across all members of the species? Or ones that are innate and instinctive and not dependent at all on maturation or experience? If those are the bars that are set, then not much gets over them.

But if instead we define human nature as a set of behavioral *capacities* or *tendencies* that are typical of our species, some of which may nevertheless be shared with other animals, and which may be expressed either innately or require maturation or experience to develop, then the list is long and much less contentious. Humans *tend to* walk upright, be active during the day, live in social groups, form relatively stable pair-bonds, rely on vision more than other senses, eat different kinds of food, and so on. A zoologist studying humans would say they are bipedal, diurnal, gregarious, monogamous, visual, and omnivorous—all of these traits are shared by some other species, but that overall profile characterizes humans.

And humans have *capacities* for highly dexterous movements, tool use, language, humor, problem solving, abstract thought, and so on. Many of those capacities are present to some degree in other animals, but they are vastly more developed in humans. The actual behaviors may only emerge with maturation and many depend to some extent on learning and experience, but the capacities themselves are inherent. Indeed, even our capacity to learn from experience is itself an innate trait. Though our intellect separates us from other animals—by enabling the development of language and culture, which shape all of our behaviors—our underlying nature is a product of evolution, the same as for any other species.

Simply put, humans have those species-general tendencies and capacities because they have human DNA. If we had chimp DNA or tiger DNA or aardvark DNA, we would behave like chimps or tigers or aardvarks. The essential nature of these different species is encoded in their genomes. Somehow, in the molecules of DNA in a fertilized egg from any of these species is a code or program of development that will produce an organism with its species-typical nature. Most importantly, that entails the specification of how the brain develops in such a way that wires in these behavioral tendencies and capacities. Human nature, thus defined, is encoded in our genomes and wired into our brains in just the same way.

This is not a metaphor. The different natures of these species arise from concrete differences in some physical properties of their brains. Differences in overall size, structural organization, connections between brain regions, layout of microcircuits, complement of cell types, neurochemistry, gene expression, and many other parameters all contribute in varied ways to the range of behavioral tendencies and capacities that characterize each species. It's all wired in there somehow. Human nature thus need not be merely an abstract philosophical topic—it is scientifically tractable. We can look, experimentally, at the details of how our species-typical properties are mediated in neural circuitry. And we can seek to uncover the nature of the genetic program that specifies the relevant parameters of these circuits.

THE WORD MADE FLESH

To understand this genetic program, it is crucial to appreciate the way in which information is encoded in our genomes and how it gets expressed. It is not like a blueprint, where a given part of the genome contains the specifications of a corresponding part of the organism. There is not, in any normal sense of the word, a representation of the final organism contained within the DNA. Just as there is no preformed homunculus curled up inside the fertilized egg, there is no simulacrum of the organism strung out along its chromosomes. What is actually encoded is a *program*—a series of developmental algorithms or operations, mediated by mindless biochemical machines, that, when carried out faithfully, will result in the emergence of a human being.

This is not a reductionist view. The DNA doesn't do any of this by itself. The information in the genome has to be decoded by a cell (the fertilized egg, in the first place), which also contains important components required to kick the whole process off. And, of course, the organism has to have an environment in which to develop, and variation in environmental factors can also affect the outcome. Indeed, one of the most important capacities encoded in the genetic program is the ability of the resultant organism to respond to the environment.

Moreover, while the information to make any given organism and to keep it organized in that way is written in its genome, there is a web of causation that extends far beyond the physical sequence of its DNA. Its genome reflects the life histories of all its ancestors and the environments in which they lived. It has the particular sequence it has because individuals carrying those specific genetic variants survived and passed on their genes, while individuals with other genetic variants did not. A full map of what causes an organism to be the way it is and behave the way it does thus extends out into the world and over vast periods of time.

However, what we are after in this book is not a full understanding of how such systems work—how all those genetically encoded components interact to produce a human being with human nature. It is something subtly but crucially different—how *variation* in the genetic program causes *variation* in the outcome. Really, that's what we've been talking about when we've been comparing different species. The *differences* between our genomes and those of chimps or tigers or aardvarks are responsible for the *differences* in our respective natures.

INDIVIDUAL DIFFERENCES

The same can be said for differences *within* species. There is extensive genetic variation across the individuals in every species. Every time the DNA is copied to make a sperm or egg cell, some errors creep in. If these new mutations don't immediately kill the resultant organism or prevent it from reproducing then they can spread through the population in subsequent generations. This leads to a buildup of genetic variation, which is the basis for variation in all kinds of traits—most obviously physical ones like height or facial morphology. (Conversely, shared profiles of genetic variants are the basis for familial *similarities* in such traits.) Some of those genetic variants affect the program of brain development or brain function in ways that contribute to differences in behavioral tendencies or capacities.

We know this is the case because we can successfully *breed for* behavioral traits in animals. When wolves were tamed, for example, or when other animals were domesticated, early humans selected animals that were less fearful, less aggressive, more docile, more submissive—perhaps the ones that came nearest to the fire or that allowed humans to approach the closest without running away. If the reason that some were tamer was the genetic differences between them, and if those ones who hung around and tagged along with human groups then mated together, this would over time enrich for genetic variants predisposing to those traits. On the other hand, if the variation was not at least partly genetic in origin then breeding together tame individuals would not increase tameness in the next generation—the trait would not be passed on.

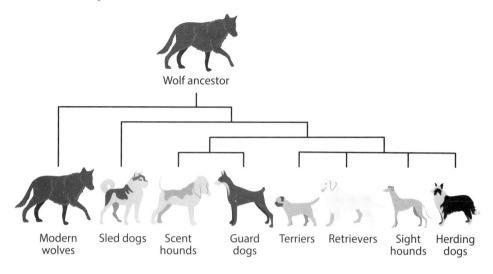

Figure 1.1 Selection of dog breeds for diverse behavioral traits.

Well, we know how that turned out—with modern dogs that have a nature very distinct from their lupine ancestors. And that process has been played out over and over again in the creation of modern dog breeds (see figure 1.1). These breeds were selected in many cases for behavioral traits, according to the functions that humans wanted them to perform. Terriers, pointers, retrievers, herders, trackers, sled dogs, guard dogs, lapdogs—all show distinct profiles of traits like affection, vigilance, aggression, playfulness, activity, obedience, dominance, loyalty, and many others. All these traits are thus demonstrably subject to genetic variation. The details of *how* genetic differences influence them remain largely mysterious, but the fact that they do is incontrovertible.

And the same is true in humans, as we will see in subsequent chapters. The empirical evidence for this is every bit as strong as it is in dogs. Even just at a theoretical level, this is what we should expect, based on the geneticist's version of Murphy's Law: anything that can vary will. The fact that our nature *as a species* is encoded in the human genome has an inevitable consequence: the natures of individual humans will differ due to differences in that genetic program. It is not a question of whether or not it does—it must. There is simply no way for natural selection to prevent that from happening.

BECOMING A PERSON

Just showing that a trait is genetic does not mean that there are genes "for that trait." Behavior arises from the function of the whole brain—with a few exceptions it is very far removed from the molecular functions of specific genes. In fact, many of the genetic variants that influence behavior do so very indirectly, through effects on how the brain develops.

This was dramatically highlighted by the results of a long-running experiment in Russia to tame foxes. Over 30 generations or more, scientists have been selecting foxes on one simple criterion—which ones allowed humans to get closest. The tamest foxes were allowed to breed together and the process repeated again in the next generation, and the next, and so on. The results have been truly remarkable—the foxes did indeed end up much more tame, but it is how that came about that is most interesting.

While they selected only for behavior, the foxes' appearance also changed in the process. They started to look more like dogs—with floppier ears and shorter snouts, for example—even the coat color changed. The morphological changes fit with the idea that what was really being selected for was retention of juvenile characteristics. Young foxes are tamer than older ones, so selecting for genetic differences that affected the extent of maturation could indirectly increase tameness, while simultaneously altering morphology to make them look more like pups.

This highlights a really important point. Just because you can select for a trait like tameness does not mean that the underlying genetic variation is affecting *genes for tameness*. The effect on tameness is both indirect and nonspecific, in that other traits were also affected. Though their identities are not yet known, the genes affected are presumably involved in development and maturation somehow.

The same kind of relationship holds in us. As we will see, the genetic variants that affect most psychological traits do so in indirect and nonspecific ways—we should not think of these as "genes for intelligence" or "genes for extraversion" or "genes for autism." It is mainly genetic variation affecting brain development that underlies innate differences in psychological traits. We are different from each other in large part because of the way our brains get wired before we are born.

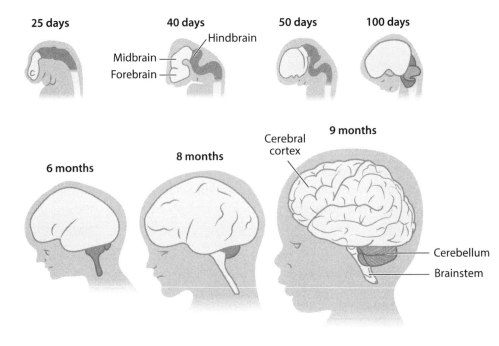

Figure 1.2 Human embryonic and fetal brain development. (Modified from B. Kolb and B. D. Fantie, "Development of the Child's Brain and Behavior," in *Handbook of Clinical Child Neuropsychology (Critical Issues in Neuropsychology)*, 3rd ed., ed. C. R. Reynolds and E. Fletcher-Janzen (New York: Springer, 2008), 19–46.)

But this is only half the story. Genetic variation is only one source of differences in how our brains get wired. The processes of development themselves introduce another crucial source of variation—one that is often overlooked. The genome does not encode a person. It encodes a program to make a human being. That potential can only be realized through the processes of development (see figure 1.2). Those processes of development are noisy, in engineering terms. They display significant levels of randomness, at a molecular level. This creates strong limits on how precisely the outcome can be controlled.

Thus, even if the genetic instructions are identical between two people, *the outcome will still differ*. Just as the faces of identical twins differ somewhat, so does the physical structure of their brains, especially at the cellular level. The progressive nature of development means that this inherent variability can have very substantial effects on the outcome, and, along with genetic differences, be a major contributor to differences in people's psychological makeup.

In sum, the way our individual brains get wired depends not just on our genetic makeup, but also on how the program of development happens to play out. This is a key point. It means that even if the variation in many of our traits is only partly genetic, this does not necessarily imply that the rest of the variation is environmental in origin or attributable to nurture—much of it may be developmental. Variation in our individual behavioral tendencies and capacities may thus be *even more innate* than genetic effects alone would suggest.

A LOOK AHEAD

This book is split into two main sections. In the first, I present a conceptual overview of the origins of innate differences in human faculties. We will start by looking at the evidence from twin and adoption studies of genetic effects on human psychological traits, brain anatomy, and brain function. These studies can begin to dissociate the effects of nature and nurture as contributors to variation across the population. They aim to explain not what makes individuals the way they are but what makes people different from each other. Because they are often misconstrued, we will look carefully at what the findings mean and what they don't mean.

We will then look in more detail at genetic variation, where it comes from and the kinds of effects it can have. We will examine how differences in the DNA sequence ultimately impact the kinds of traits we are interested in—often, as discussed above, through effects on development. We will look in depth at the mechanisms underlying the self-assembly of the brain's circuitry to see how it is affected by variation in the genetic instructions. And we will consider just how noisy and inherently variable those developmental processes can be. In the end, I hope to have convinced you that both genetic and developmental variation contribute to innate differences in people's natures.

In the final chapter of the first section we will look at the role of nurture in shaping people's psyches. The human brain continues to mature and develop over decades, and our brains are literally shaped by the experiences we have over that period. It is common to view "nurture" as being in opposition to nature, such that the environment or our experiences act as a great leveler, to smooth over innate differences between people or counteract innate traits in individuals. I will describe an alternative model: that the environments and experiences we each have and the way our brains react to them are largely *driven by our innate traits*. Due to the self-organizing nature of the processes involved, the effects of experience therefore typically act to *amplify* rather than counteract innate differences.

With that broad framework in place, we will then examine a number of specific domains of human psychology in the second section. These include personality, perception, intelligence, and sexuality. These diverse traits affect our lives in different ways and genetic variation that influences them is therefore treated very differently by natural selection. As a result, their underlying genetic architecture—the types and number and frequency of mutations that contribute to them—can be quite different. Much of the variation in these traits is developmental in origin—the circuits underlying these functions work differently in part at least because they were put together differently. This means that random variation in developmental processes, in addition to genetic variation, also makes an important—sometimes crucial—contribution to innate differences in these faculties.

We will also look at the genetics of common neurodevelopmental disorders, such as autism, epilepsy, and schizophrenia. There has been great progress in recent years in dissecting the genetics of these conditions, with results that are fundamentally changing the way we think about them. Genetic studies clearly show that each of these labels really refers to a large collection of distinct genetic conditions. Moreover, while these disorders have long been thought to be distinct, the genetic findings reveal the opposite—these are all possible manifestations of mutations in the same genes, which impair any of a broad range of processes in neural development.

The final chapter will consider the social, ethical, and philosophical implications of the framework I've described. If people really have large innate differences in the way their brains and minds work, what does that mean for education and employment policies? What does it mean for free will and legal responsibility? Does it necessarily imply that our traits are fixed and immutable? What are the prospects for genetic prediction of psychological traits? What limits does developmental variation place on such predictions? And, finally, how does this view of the inherent diversity of our minds and our subjective experiences influence our understanding of the human condition?

CHAPTER 2

VARIATIONS ON A THEME

If the typical nature of a species is written in its genome, then individual members of the species may differ in their natures due to genetic variation in that program. We saw some of the evidence for that in other animals in the previous chapter, but what about in humans? What kind of evidence could we use to determine whether genetic differences between people contribute to general differences in psychological traits? Well, one powerful method is to flip the question around and ask whether people who are more genetically *similar* to each other are also more similar in psychological traits. In short, if such traits are even partly genetic, then people should resemble their relatives, not just physically, but also psychologically.

That is a nice idea, but there is an obvious problem—people who are closely related to each other—like siblings, for example—also typically share similar environments, like being raised in the same family. So, if we know only that siblings resemble each other psychologically more than random people in the population, we cannot distinguish possible effects of nature from those of nurture. We need some way to dissociate these two effects—to test the impact of shared genes separately from the impact of shared family environment, and vice versa.

TWIN AND ADOPTION STUDIES

Twin and adoption studies have been developed for precisely that purpose. Adoption studies are the simplest to understand—the idea is that if shared genes are what make people more similar to each other, then adoptees will resemble their biological relatives, while if shared environment is more important then they will resemble their adoptive relatives, especially adoptive siblings (children who are not biologically related but who are raised in the same family).

Twin studies take the converse approach—they compare people who have the same degree of shared environment, but differ in how similar they are genetically. Twins can be identical (or monozygotic [MZ], meaning they come from a single fertilized egg, or zygote, that has split into two embryos with the same genome) or they can be fraternal (or dizygotic [DZ], meaning they come from two different eggs fertilized by two different sperm and thus are only as similar to each other as ordinary siblings—they just happen to be conceived at the same time). As they grow up under similar conditions, these different types of twins make an ideal comparison to test the importance of shared genes.

If the environment you grow up in were the only thing that mattered for some trait, then the similarity between pairs of MZ twins should be about equal to that between pairs of DZ twins. DZ twins make the ideal comparison here because they grow up not just in the same household, but at the same time, and also share any possible effects of being twins, which, if they exist, would not be apparent in other siblings. By contrast, if variation in a trait is due to genetic differences, then MZ twins should be more similar to each other than DZ twins. Of course that is obviously true for physical traits, which is why we call MZ twins "identical." But is it true for psychological traits?

To answer this question, we need to do something that is much harder for psychological traits than for physical ones like height—we need to measure them. If we are to calculate how similar different people are for some trait, we need a number—some objective measure that captures or reflects variation in the trait of interest.

MEASURING PSYCHOLOGICAL TRAITS

There are many possible ways to do this, some of which are more direct than others. For example, we can simply ask people questions about their own behavioral patterns or predispositions and generate some kind of arbitrary numerical ranking or score from their answers, as in personality questionnaires. These typically ask people how strongly they agree or disagree with statements like "I really enjoy going to parties and get energized by social situations," and give a score based on a five-point scale. If you analyze the responses to many such questions you can get an aggregate number that reflects the personality trait of extraversion.

These kinds of questionnaires were first developed by Francis Galton, the Victorian polymath, who was obsessed with measuring anything that could be measured, and who applied this to the study of variation in human faculties. He also devised ways of classifying fingerprints, created the first weather map, and even studied scientifically the best way to make a cup of tea. It was Galton who coined the phrase "nature versus nurture" and he foresaw the use of twin and adoption studies as a means to separate these effects. Later, he became a champion of the eugenics movement (having invented the term), which led to a dark chapter in the history of human genetics, not just with the well-known horrors in Nazi Germany, but also with the enthusiastic adoption of eugenic policies in the United Kingdom and the United States, involving forced sterilizations of "feeble-minded" people. Though the days of enforced government programs such as this are hopefully over, new genetic technologies are providing the means for individual action, in selection of embryos based on genetic information, for example. This raises a host of ethical and moral issues, which we will consider in chapter 11. In the meantime, we will see more of Mr. Galton in this and subsequent chapters.

An alternative to questionnaires is to measure performance on tests of, for example, intelligence or memory or empathizing—anything where a specific number emerges based on success in answering questions. This can be extended to all kinds of tasks in a lab where things like reaction time or quantitative differences in perception or task performance are measured. And these days we can go even further and directly measure differences in brain structures or brain activity under various conditions and consider such differences as traits of interest.

Finally, we can measure the actual occurrences of specific behaviors or of real-world outcomes that can act in some way as proxies for underlying traits. These might include things like educational attainment, number of times arrested, what time you get up in the morning, number of same-sex partners, whether you have ever been prescribed an antipsychotic medication, how much you drink, whether you write with your right or left hand, and so on.

With all these methods, the important thing is that we get a number for each person that we can use to then ask how similar or different people are. I should emphasize here that the use of such measurements is not the same as "reducing a complex behavior to a single number," as is sometimes charged. They are simply experimental tools that allow us to ask some interesting questions. This kind of methodological reductionism is merely aimed at making complex questions tractable by defining measurable parameters that allow precise experimental questions to be formulated and tested. It does not constitute a philosophical commitment to theoretical reductionism—the idea that complex behaviors relate to such simple measures in a relatively straightforward fashion. They clearly do not, but that should not stop us from asking and answering some interesting questions about the factors that contribute to complex behaviors.

That said, these measures are clearly a lot fuzzier and less exact than measures of traits like height or weight. Indeed, we might be concerned that they don't measure anything real at all—that they are simply noise. That is clearly not the case. We can in fact measure how good our measurements are by testing people numerous times and seeing how consistent the results are. If I took a personality questionnaire one day and it said I am highly extraverted and I took it again a week later and it told me I am very shy and reserved, well then I would say that test is not very reliable or informative. Or if my IQ varied wildly over different test sessions, I would reject it as a useful measure. In fact, a huge amount of effort has gone in to creating questionnaires, tests, and tasks that do have high test-retest reliability, generating highly consistent measurements within individuals. Note that the question of what such measurements *mean* is a separate one—one that we will get into in subsequent chapters. For now, it is enough to know that they are actually measuring *something*, a real thing that exists—a trait that differs between people. Geneticists call that the *phenotype*—the outward manifestation of some underlying difference.

COMPARING TRAITS ACROSS PEOPLE

Now that we have some measurements related to our traits or phenotypes of interest, we can get back to the idea of comparing people to see how similar they are. What we want to do is get an estimate of similarity

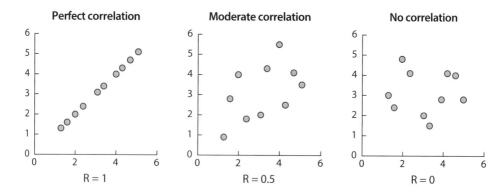

Figure 2.1 Correlations. Plots of values of a trait for twin 1 versus twin 2, across many such pairs. If the values are identical, the correlation coefficient, R, will be 1. If they have no relationship, R will be 0. Intermediate values indicate a partial correlation.

not just in one pair of individuals, but across large sets of pairs of individuals of different types. That might be across many pairs of adopted or biological siblings, or many pairs of MZ or DZ twins. One way to visualize these relationships is to draw a graph, with the values for one person in a pair on one axis and the values for the other person in each pair on the other axis. If we think about height in twins, for example, then if one twin in a pair is 5′8″ tall and the other is 5′9″ tall, we place a point at the intersection of those coordinates on the x and y axes (as in figure 2.1). Now our next set of twins might be 6′2″ and 6′3″ and we would plot another point at those coordinates.

If we keep doing that we will get a visual picture of how similar our twins are to each other. If within each pair they are identical in the trait being measured, then all the points will fall on a straight diagonal line. If, on the other hand, there is no similarity within pairs (as would be seen if we just take random subjects from the population and assign them to pairs), then the dots will be scattered randomly all over the place. And if the individuals within pairs are more similar to each other than random strangers, but not quite identical, then the dots will fall generally near the diagonal line, but will be scattered a bit around it.

Those graphs give a very nice intuitive representation of the strength of the relationship within pairs, but we can go further than that and mathematically calculate a number that precisely describes that relationship, which is called the *correlation* or *regression coefficient* (another invention of Francis Galton). This number ranges from 0 (if there is no relationship within pairs) to 1 (if the values are always identical within pairs).

If we make such plots and calculations for many pairs of adopted siblings, we tend to find a very modest correlation between them for many psychological traits. They are, for some traits, more similar to each other than random people, but typically only slightly and in many cases such similarity seems to be temporary—it is evident if the trait is measured while the siblings still live in the same home, but tends to disappear if they are measured as older adults. By contrast, if we plot adoptees versus their biological siblings, we see a much stronger correlation for many such traits. These findings indicate that sharing genes with other people really does make you more similar to them psychologically, and that this effect is not due to having similar upbringing. In fact, the effect of a shared family environment is remarkably modest for most such traits.

We can do a similar comparison between MZ and DZ twins. Typically what is found is that MZ twins are much more similar to each other than are pairs of DZ twins. As they share a family environment similarly in each case, this effect must be due to the fact that MZ twins share all their genetic material, while DZ twins share only 50% of it. Indeed, for many traits, MZ twins who have been reared apart are just about as similar to each other as ones who have been reared together. Again, the conclusion is that shared genes have a much bigger effect on psychological traits than a shared upbringing.

These data directly show that people who are more genetically similar to each other tend to be more phenotypically similar to each other for psychological traits, and this correspondence is not due to being raised in the same family. From that, we can draw a more general inference by flipping back to considering differences rather than similarities. We can infer that genetic *differences* between people make a big contribution to *differences* in psychological traits across the population. By contrast, differences in family environments make a much smaller, often negligible, contribution. Now we are thinking not about what makes one individual a certain way or what makes two individuals similar to each other—we are instead thinking about what factors contribute to variation in a trait across the whole population. It is worth pausing a moment to consider what that shift in perspective means.

VARIATION ACROSS THE POPULATION

If we measure a trait in many individuals in the population, then we will see some variation in that trait and we can measure that too. For traits that have a continuous range of values, like height or IQ, we can plot the distribution of values across the population in what is known as a histogram. A histogram plots the values of the trait along the horizontal axis and the number of people who have that value on the vertical axis. Generally speaking, you find many more people near the average value and far fewer people as you go out to the extremes, giving the famous bell-shaped curve, or "normal distribution" (see figure 2.2).

All normal distributions have that general bell shape, but for some the bell is higher and narrower and for others it is lower and wider. A low and wide curve shows that there is more variability in that trait across the population. For example, if you plot the heights of all males in the population, you will have a distribution that ranges from well under five feet to over seven feet, with many more people in the middle of the range than at the ends. But if you were to plot the heights of professional basketball players, you would get a distribution with the average height shifted far toward the higher end and with much less variability—the bell curve would be narrower. (You can imagine a similar situation at the other end of the spectrum if you were to plot the heights of jockeys.)

The amount of variability seen in a distribution of values like that is called the *variance*. The variance is a precise number—it is calculated by measuring how far each point is from the mean, or average point, of the distribution, then squaring these values (so that any difference becomes a positive number) and adding them up. So, if the values are all clustered very close to the mean, the variance will be a small number. But if the values range more widely, the variance will be larger. And that is what we want to explain. What is it that causes that variability? What are the sources of variance in the trait?

Twin and adoption studies allow us to estimate how much of the variance in a trait is due to genetic differences between people, how much is due to differences in family environments, and, importantly, how much is unexplained even when those two factors are taken into account. For example, if we find that MZ twins are much more similar to each other

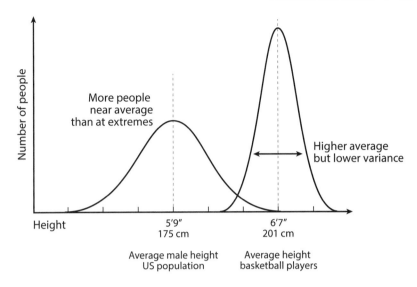

Figure 2.2 Variance. The distribution of heights across the general population follows a normal bell-shaped curve. The width of this curve reflects the variance of the trait. A selected population, such as professional basketball players, shows a higher mean value, but lower variance.

than DZ twins are, then we can infer that genetic differences make a large contribution to the variance in the trait across the population. Alternatively, if MZ twins are just a little more similar than DZ twins, then genetic differences must not play as large a role. Or if we find that adoptive siblings are just as similar to each other as biological siblings for some trait, that implies that a shared family environment is the key factor—that is, that differences in family environments can fully explain the variance in the trait that we observe across the population.

It's possible to go beyond just general statements, though. By mathematically comparing the values of the correlations between these different sets of pairs (MZ vs. DZ twins, or adoptive vs. biological siblings, or many other possible combinations), we can calculate the *percentage of the variance* accounted for by these various factors. The results from hundreds of such studies are remarkably consistent. In general, for psychological and behavioral traits, the percentage of variance accounted for by genetic differences ranges from modest (30%–40%) to very high (70%–80%). This last factor—the amount of variance in a trait that can be attributed to genetic variation—is known as the *heritability*. It is a key concept in genetics but one that is often misunderstood—more on what it means and doesn't mean below.

Importantly, heritability can also be estimated by comparing phenotypes across thousands of people in the general population, relying on the fact that we are all distantly related, to varying extent. Even a small increase in relatedness above the average level is enough to cause a slight, but measurable, increase in phenotypic similarity. This effect can be measured by carrying out millions of pairwise comparisons across a population sample of thousands of people. Results from these kinds of studies confirm the heritability of psychological traits and also demonstrate that they are not caused by any artifacts or unusual aspects of twin studies.

These findings hold for all kinds of personality traits—conscientiousness, extraversion, impulsivity, aggressiveness, threat sensitivity, warmth, and on and on—as well as intelligence, memory, language ability, motor skill, balance, psychological interests, sexual orientation, sleep patterns, musicality, appetite, social attitudes, even how religious people are. It holds for behaviors like smoking, problem drinking, antisocial behavior, educational attainment, marital fidelity, and likelihood of divorce. And it is true for all kinds of psychiatric disorders, anxiety, drug abuse, even suicidal behavior. For all these traits or behaviors, genetic differences are a substantial cause of variation across the population. Within families, MZ twins are much more similar to each other than DZ twins, and biological siblings are much more similar to each other than adoptive siblings.

The clear biological basis for these traits suggests that twins and siblings behave similarly to each other because their brains are wired in similar ways. With new neuroimaging technologies we can now see that that is literally true.

VARIATION IN BRAIN STRUCTURE

Neuroimaging techniques allow us to see the structure of the brain in ever-increasing detail, and also to visualize its activity under different conditions. The most prominent of these techniques is magnetic resonance imaging (MRI). MRI works by using strong magnetic fields to alter the states of atoms in a tissue, particularly hydrogen atoms. The single proton in the nucleus of each hydrogen atom acts like a tiny compass needle and aligns to the magnetic field. Radio waves are used to knock these protons out of alignment, and as they relax back into alignment they give off a radio wave signal that can be detected from outside the body. By using a graded and pulsed magnetic field, the radio wave signals can be localized with very high precision to create a high-resolution, three-dimensional scan of the tissue. Differences in these signals are caused as hydrogen atoms in different tissues realign at different speeds. This generates a contrast between different regions of a tissue—for example, between muscle and bone and tendon in your shoulder or your knee, or between gray and white matter in your brain.

"Gray matter" refers to areas of the brain where nerve cell bodies are densely packed and where the thin fibers that connect them (called *dendrites* and *axons*) are diffuse and local, intermingled among the cells. "White matter," on the other hand, refers to large bundles of axons that run quite separately from the nerve cell bodies and connect distant regions of the brain. They appear white because they are insulated with a fatty substance called myelin. The white matter is organized into major pathways through the brain—connecting, for example, the two cerebral hemispheres, the front of the brain to the back, the outer cortex to lower areas, or the brain to the spinal cord.

MRI scans can provide extremely detailed three-dimensional pictures of an individual's brain, from which we can extract all kinds of measurements. The most obvious is the volume or thickness or surface area of various brain regions, or the volume of specific white matter tracts. With these scans and measurements in hand, we can ask whether people who are genetically more similar to each other have brains that are structurally more similar. Outwardly, MZ twins look "identical" to each other—we can now clearly see that this similarity extends to the physical structure of their brains. Indeed, the brain scans of the two MZ twins shown in figure 2.3 look at first like scans of the same brain. Closer inspection reveals some subtle differences, but they are clearly much more similar to each other than are the brains of the DZ twins or siblings also shown. Importantly, the measurements we can derive from these scans allow us to quantify this similarity across large numbers of pairs of twins and assess their heritability, exactly as we did for psychological traits.

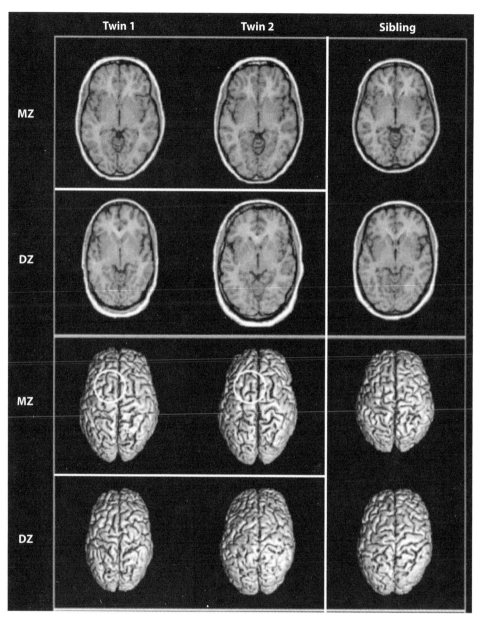

Figure 2.3 Twin brains. MRI scans show that the structure of the brain is almost identical between monozygotic (MZ) twins—much more similar than dizygotic (DZ) twins or nontwin siblings. White circles highlight subtle differences in brain folding patterns between MZ twins. (Reprinted from A. G. Jansen, S. E. Mous, T. White, D. Posthuma, and T. J. Polderman, "What Twin Studies Tell Us about the Heritability of Brain Development, Morphology, and Function: A Review," *Neuropsychol. Rev.* 25 (2015): 27.)

The results from these kinds of twin studies are striking and bear out the impression given by figure 2.3. Many aspects of the structure of the brain are very highly heritable—that is, most of the variation is due to genetic differences across individuals. Heritability estimates for various measures are: total brain volume, 82%; gray matter volume, 72%; white matter volume, 85%. The heritability for volumes of particular parts of the cortex or of other brain structures ranges from 60% to 80%, while measurements of the thickness of various parts of the cortex show heritabilities of around 50%–70%. Similar results are seen even in very young infants, only a month old.

MRI data can also be used to look at brain organization—specifically, how different parts are connected to each other. A technique called diffusion-weighted imaging tracks the direction of diffusion of water molecules inside the brain and can detect the orientation of bundles of nerve fibers. Using these signals it is possible to measure the extent of nerve fiber connectivity between different areas and even to extract information on the overall brain network. Twin studies of these kinds of measures reveal moderate to very high heritability of local measures of the size or microstructural organization of individual tracts.

It is also possible to use measures of specific tracts to build a picture of the connectivity of the entire network of brain regions. These networks can then be analyzed to reveal subnetworks of interconnected regions, and characterized mathematically to derive various measures that describe the overall patterns of connectivity, including how clustered the connections are and how efficient information flow will be through the network. These kinds of network parameters also show heritabilities in the range of 60%–70%.

Collectively, these data show that much of the physical variation in brain structure between individuals is attributable to genetic differences. To put it more simply, our genes have a big effect—by far the predominant effect, in fact—on how our brains are wired, very literally. These findings bear on a common misconception about the role of genes—it does not end at birth. It is not the case that genes establish the initial brain wiring pattern and everything else depends on experience. The genetic program of brain development entails all the growth and maturation that occurs after birth, exactly as for other parts of the body.

VARIATION IN BRAIN FUNCTION

Neuroimaging techniques can show us not just how people's brains are wired, but also how they work, at least at a gross level. A technique called functional magnetic resonance imaging (fMRI) is a powerful method that allows us to see which areas of the brain are active. It relies on the fact that active areas of the brain attract a flow of oxygenated blood, which has a different magnetic resonance signature to deoxygenated blood. Though this signal is much slower than the neuronal activity itself, it is a reliable proxy for that activity over a time frame of several seconds. This is the technique that is widely used to track which parts of the brain are involved in various functions. When you read about areas of the brain "lighting up" when you see a rattlesnake or hear music or think of serving a tennis ball, this is the signal they're talking about.

In reality, it relies on a lot of unglamorous statistical analysis to extract the signal from both the noise and the background activity. This raises a crucial point: though parts of the brain can be "activated" by various stimuli, this does not mean they are normally just sitting there, not doing anything. The brain is always active, even when a person is at rest—or even asleep, for that matter—a bit like a car sitting with its engine running, just idling.

That idling activity can also be detected by fMRI, and one of the things that people have noticed is that different parts of the brain sit there humming along at different frequencies. The fMRI signal shows a slow fluctuation or oscillation in each area, becoming slightly stronger or slightly weaker every 10–20 seconds or so. If you just let a person rest in the MRI scanner for about five minutes, you can track these fluctuations across all the areas of the brain. Then you can do something really interesting—you can see which areas of the brain are fluctuating *in synchrony* with each other.

When a person is engaged in some task, different parts of the brain may be coactivated. These usually reflect brain regions that make up an extended circuit or system involved in whatever that task is. It turns out that those functional relationships are also evident in the temporal correlations of the spontaneous fluctuations at rest. These are thought to reflect a past history of coactivation, meaning that if two areas are fluctuating in synchrony with each other, they are likely part of an extended functional network. Importantly, while there is a general pattern to the subnetworks that emerge through these kinds of analyses, there are also important individual differences. Repeated imaging of the same people shows that such differences are highly reliable, reflecting stable differences in functional brain architecture, which are also highly predictive of the pattern of activity during various tasks across individuals. Indeed, these networks are so distinctive that they provide a kind of "neural fingerprint" that can be used to reliably identify individuals from brain scans, regardless of what the brain is actually engaged in during imaging.

Moreover, since the degree of temporal correlation gives a quantitative measure of the strength of functional connectivity between any two brain areas, these correlations can be used to derive a brain-wide *functional* connectivity network, just as for structural parameters. Structural and functional connectivity networks generally show very good correspondence, though many areas may be functionally "connected"—that is, talking to each other—even if they do not share a direct structural connection. And, again, multiple parameters of these networks can be measured and compared between people, including pairs of twins. The result, which is unlikely to surprise you at this stage, is that the brain networks of MZ twins are much more similar to each other than are those of DZ twins, such that both local and global parameters of functional connectivity show moderate to high heritability.

The upshot of all this is that the brains of people who are genetically related to each other are wired similarly and work similarly. Presumably this underlies their similarities in psychological traits. Once again, if we flip perspective, we can infer that a substantial proportion—often a majority—of the variance in the population in both brain traits and psychological traits is due to genetic differences. Now, it is important to delve a little deeper into this concept of heritability.

HERITABILITY—WHAT IT MEANS AND WHAT IT DOESN'T MEAN

One of the crucial things to keep in mind about heritability estimates is that they refer to *variance*, not to mean or absolute values of a trait. We use them to understand what makes people different from each other, or different from the average value in a population—they say nothing about why that average value is what it is. That question comes back to our discussion about species-general traits; what we want to understand here is variation around those mean values, within a species. What drives the mean is still genetic in the sense that it still depends on our genomes—it's just not what we're interested in here. We are all generally human sized because of our human genomes—what heritability estimates are relevant to is the question of what makes some humans taller or shorter than others.

So, if we find that the heritability of a trait is, say, 60%, this does not mean that 60% of the absolute value of that trait in a particular individual comes from his or her genes. It would make no sense to say 60% of my height is genetic, for example. It means that, *across the population*, 60% of the variance (the deviation of individuals from the mean value of the trait) is due to genetic differences. So, if everyone in the population were genetically identical, the variance in the trait would be only 40% of what it actually is.

This brings up another crucial fact about heritability—it is a *proportional* measure. Say we have some trait that can be affected by both genetics and environment. Height is a good example, as there are strong genetic effects on a person's potential final adult height, but whether that height is actually attained can be affected by nutrition. If we measure heritability of height in a population where everyone has ready access to food, it will likely be quite high. Most of the variance in the trait will be due to genetic differences, partly because there are few differences in other factors that matter. But if we measure it in a population where access to food is highly unequal, then we may find the heritability is lower. This doesn't mean the genetic effects have been reduced in an absolute sense—just that their *relative* importance to the overall variation in the trait is lower, because the environmental variance is higher. Because of this, heritability estimates are always local and historical, applying only to the population in which the trait was measured. The number we find in any study is not a biological constant, equivalent to those we find in physics. It doesn't measure what factors *can* affect a trait; it only measures what factors actually *do* affect a trait, in a given population at a given time. The environment can still affect the mean value of a trait, but if it doesn't vary much then it won't contribute to *differences* in the trait.

Because heritability tells us only about sources of variance *within* a population and nothing about why the mean value is what it is, it also tells us nothing about sources of differences in mean values *between* populations. It is quite possible to have a trait that is highly heritable in two populations, but where the difference in the mean value between the populations is caused by nongenetic factors. Body mass index (a measure of weight relative to height) is a good example of this. It is highly heritable when measured within individual populations, but a comparison across countries shows huge disparity in average body mass index and percentage of the population that is overweight or obese. These differences are not genetic in origin; they are environmental or cultural. This issue is especially important when it comes to interpreting the heritability of intelligence and the possible causes of differences in average IQ across populations or over time. We will see in chapter 8 that an exactly analogous situation holds for intelligence as for body mass index.

Finally, it is important to emphasize that heritability is not the same as heredity or inheritance, or at least not always. For animal breeders, heredity is the important aspect—how strongly offspring resemble their parents. But heritability actually measures all genetic influences on a trait, not all of which are actually *inherited*. First, many traits are caused by multiple genetic factors acting together—the particular combinations of genetic factors may be crucial in determining the phenotypic outcome in each individual. Because each of our genomes represents a new combination of those genetic variants, these will be different from either of our parents. Second, we each also have new mutations in our genomes that arose in the generation of the sperm and egg cells from which we were formed. These also contribute to our individual traits but obviously do not contribute to parent-offspring similarity. Down syndrome provides a stark example of this; it is a condition that is rarely inherited from a parent—it most often derives from a new event in the egg or sperm that leads to an extra chromosome 21 being included—but it nevertheless has a completely genetic mechanism in the individual. Both these factors—the influence of new mutations and the importance of unique combinations of genetic variants—make large contributions in twin studies as MZ twins share all new mutations and also the exact same combinations of all genes.

NONGENETIC EFFECTS

I have been emphasizing the heritability of psychological and brain traits in humans, but twin and adoption studies also highlight *nongenetic* contributions to overall variance. These effects are often assumed to be "environmental" in origin, but we will see that that is not necessarily the case. The same comparisons of MZ and DZ twin pairs or biological versus adoptive siblings that are used to calculate heritability can also be used to estimate the variance explained by different family environments.

Consistently, and surprisingly, this turns out to be very low (usually not more than 10%–15%) and is often found to be zero. Generally speaking, adoptive siblings do not resemble each other for psychological traits any more than two strangers in the street. This is despite being raised in the same household, living in the same community, typically attending the same schools, and so forth. And for many traits, MZ twins who are reared apart are almost as similar to each other as those who have been reared together—sharing a family environment does not make them appreciably more similar.

This result has caused some consternation and even disbelief over the years since it was first highlighted by, for example, Judith Rich Harris and Steven Pinker. However, it is actually far less surprising if we consider the *kinds of traits* we are talking about. They are the very ones that, by definition, reflect some stable differences between people, some underlying dispositions that influence patterns of behavior over time. Any parent with more than one child will likely have noticed differences between them that cannot be traced to differences in parenting—in fact, these are an endless topic of conversation between parents. Why is one child studious and attentive while the other has his or her head in the clouds? Why is one cautious while the other is on a first-name basis with staff at the emergency room? Why is one so shy and quiet that you worry he or she will never have any friends while the other would happily stand talking to a post? Children have different temperaments, different talents, and different interests that simply seem to emerge of their own accord and to be largely resistant to any efforts to change them.

Academics love to find things that are counterintuitive—that conflict with our everyday experience and show how wrong we can be about the way our minds work. This is not one of those cases. The results from twin studies do not actually conflict with our intuitions and our common experience at all. These studies are about precisely those kinds of traits that we encounter as parents as largely innate or intrinsic to the child. We should not be surprised if the results fit with this experience.

Does this mean parenting doesn't matter? Of course not. It doesn't even mean that parenting doesn't affect our offspring's behavior—of course it does. Love, encouragement, support, discipline, expectations: all have hugely important impacts on children's lives. They shape the characteristic adaptations we all have to the situations in our lives, to our expectations of ourselves and the choices we make. It just means that parenting doesn't significantly affect their underlying behavioral *traits* or predispositions. But those traits are only part of what influences people's actual behavior.

These findings suggest that many reported correlations between parental behavior and offspring traits do not reflect a direct causal link, as often inferred, but instead reflect the effects of shared genes. If, for example, we find that overprotective parents have anxious children, this could be because overprotective parenting *causes* children to be anxious. But the evidence described above is not consistent with such an interpretation, as it should affect MZ and DZ twins or adoptive and biological offspring equally. Instead, the general findings suggest that parental overprotectiveness and child anxiety are more likely both manifestations of the same genetic effects, acting in both the parents and the offspring. Similarly, growing up in a household with more books in it is correlated with higher IQ—does this mean reading raises your IQ? Well, I'm all for reading, but this correlation more likely reflects the fact that parents with higher IQ tend to have more books in the house and also tend to have children with higher IQ. In general, these kinds of sociological correlations are thus hopelessly confounded by possible (indeed, likely) genetic effects.

It should be stressed, however, that most twin and adoption studies sample a relatively small range of potential family environments. Many studies have shown that serious neglect or abuse can have long-lasting psychological consequences. Fortunately, such situations are rare, at least rare enough that they do not make much of a contribution to overall variance in psychological traits across the population and thus do not show up in the shared family environment component of variance. Again, these studies only measure the factors that actually do make a contribution to variance in a population—not all the ones that could make a difference, if they occurred.

If genetic effects account for 40%–60% of the phenotypic variance and family environments account for only 0%–10%, that clearly leaves a good chunk of the variance unexplained. Something else is making people different from each other, even MZ twins who grow up in the same family. That factor is referred to as the "nonshared environment," but we will see in later chapters that much of this may be caused not by any factors outside the organism, but by inherent variation in the processes of development themselves.

In the next chapter, however, we will concentrate on the genetic effects. We will look at what genes actually are, where genetic variation comes from, and how it affects the kinds of traits we are interested in.

CHAPTER 3

THE DIFFERENCES THAT MAKE A DIFFERENCE

When we say that genes influence behavior, what we really mean is that genetic *differences* contribute to *differences* in behavioral traits (which in turn influence patterns of behavior over time). So, what are these genetic "differences"? To answer that, we need to start with a more basic question: What are genes?

You might think there is a simple answer to that question, but there isn't. Defining what a gene is has in fact been a source of enormous confusion both within science and for the general public. The reason is that the term actually refers to two very different things. The original concept, famously devised by Gregor Mendel in the 1850s in studying various traits in peas, was of some physical thing that gets passed on from parent to offspring, and that determines the trait in question. From the patterns of inheritance he inferred that there must be distinct genes for whether peas had smooth or wrinkled shells, whether they were green or yellow, whether the flowers were white or purple, whether the plants were tall or short. He also was able to deduce that each plant inherited two copies of each gene—one from the mother and one from the father. Importantly, Mendel realized that each of these traits was controlled by a discrete inherited unit—different ones for different traits. The term "gene" was introduced later to refer to these units of heredity.

While Mendel knew that these units must have some physical substrate—genes must be a physical thing—he didn't know what they were made of. It was not until the 1940s that scientists figured out that the genetic material was DNA—deoxyribonucleic acid, a major chemical constituent of the chromosomes (literally, colored bodies) that were visible down the microscope in the nucleus of cells.

This fact is so well known now that it's hard to think of a time when it wasn't, but actually DNA was not even a front-runner in the betting for what substance carried the genetic information. It was deemed too simple, as it is composed of only four different chemical subunits, or bases, arranged in a long sequence along each chromosome. The preferred candidate was proteins, also present in chromosomes and throughout cells— these are much more complicated than DNA, as they are composed of 20 different amino acids strung together in long chains, which then fold back on themselves to form complicated three-dimensional shapes. While DNA just kind of sits there, proteins are properly impressive—they do all sorts of things inside cells, acting like tiny molecular machines or robots, carrying out tens of thousands of different functions.

Proteins thus seemed a much likelier candidate than DNA to be the genetic material. But a seminal experiment looking at how one type of bacterium could be transformed from a nonvirulent to a virulent (disease-causing) form clearly showed that it was DNA and not proteins that carried this genetic information. (It turns out the proteins associated with chromosomes are involved in packaging the DNA inside cells and in regulating which genes are expressed, but do not themselves carry the genetic information.) From our vantage point, in the digital age, this now seems unsurprising. The simplicity of DNA that led many to dismiss its information-carrying capacity can now be seen as ideal if the information is carried in the *sequence* of the bases that make it up, just as it is carried in a sequence of 1s and 0s in a computer. Moreover, the fact that it is chemically very inert—it just doesn't do much—is exactly what you want in order to safely and stably encode information over long periods, not just over the lifetime of an organism, but also over many generations spanning millions of years.

In fact, these properties were predicted on theoretical grounds by the physicist Erwin Schrödinger in a famous series of lectures on "What Is Life?" delivered in 1943 at Trinity College Dublin. He realized that what set living things apart from nonliving ones was that living things are organized. Both living and nonliving things are made of the same kinds of stuff—of atoms—it's just that in living things these atoms are organized into molecules and complexes of molecules and cells and organs. Keeping things organized is hard work, as the general trend in the universe is for things to get messier, if left to their own devices. It requires energy to keep things organized, which all living things must take in, in some form or another, but it also clearly requires information. An organism must contain within it the information for *how* all those molecules and cells should be organized. And it must be able to replicate that information and pass it on to its offspring. Schrödinger realized that what he called an "aperiodic crystal" would be a perfect medium to store such information—that is, the material should be stable, like a crystal, and should contain within its structure a code, written in the nonrandom, nonrepeating sequence of different subunits.

THE STUFF THAT GENES ARE MADE OF

DNA fits that bill perfectly. The most obvious and direct thing encoded in DNA is, a little ironically, proteins. The recipes for all those impressive micromachines whizzing around in our cells are written in our DNA. And this brings us to the second definition of a gene—one derived from molecular biology, rather than the study of heredity. Here, a gene is a stretch of DNA that codes for a specific protein. Each chromosome in the cell is a single continuous molecule of DNA, like a long string, made of a series of the four different chemical subunits joined together. These subunits are called adenine, thymine, cytosine, and guanine, but are usually referred to as the "letters" of the DNA code: A, T, C, and G, respectively. Each of these molecules has a polarity to it—they have two ends where they can be chemically joined with the other bases—actually rather like the way we join letters together to make words.

The chromosome is made of two of these strands of DNA wound around each other in the iconic double helix. The information on each strand is complementary to the other due to the way that the chemical bases interact with each other: an A on one strand will be matched by a T on the other, while a C on one strand will pair with a G on the other. This gives an obvious mechanism for copying DNA—the double helix can be unwound and the two strands pulled apart, with each one then acting as a template for construction of another version of the other one, yielding two copies of the double helical molecule.

If you start at one end of a chromosome and scan along it (on one strand), you will soon come to a bit of the DNA that is special, because

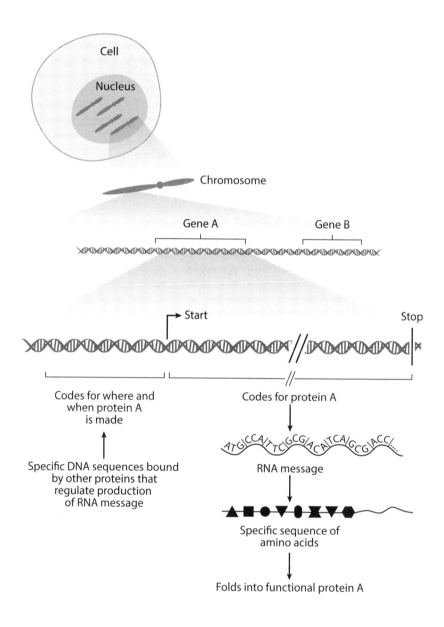

Figure 3.1 The physical structure of genes. Spread out along each chromosome are genes—stretches of DNA that code for proteins. The sequence of DNA bases—A, C, G, and T—codes for a sequence of amino acids in the corresponding protein. DNA sequence in regulatory regions controls protein expression.

the sequence of bases here encodes a protein. That is, the sequence of letters is a code that tells the cell which amino acids to string together, in what order, to make protein A or protein B, and so on. It took a while to work out, but we now know that each successive three-letter stretch of the DNA sequence corresponds to a different amino acid. There are also three-letter codes that tell the cell where the code for a particular protein starts and where it ends. So, if you keep scanning along the DNA, you will also come to the end of the section that codes for whatever that protein is. Figure 3.1 illustrates the structure of a gene.

From a molecular biological point of view—the perspective that aims to understand how cells work rather than how traits are inherited—that stretch of DNA is a "gene." We have about 20,000 different genes spaced out along our 23 chromosomes, which collectively make up the human genome. They code for proteins like collagen, hemoglobin, insulin, metabolic enzymes, antibodies, ion channels, neurotransmitter receptors—all the things that cells need to do their various jobs.

TURNING GENES ON AND OFF

Now, things are about to get more complicated. When I said that a gene encodes a protein, that is true, but the gene itself doesn't make the protein. As I mentioned above, DNA is an incredibly inert molecule—it just stores the information. In order for that information to be acted upon, or expressed, it must be read out by the cell and decoded. The machinery that does that is itself composed of other proteins in the cell. (If you're starting to see a chicken and egg problem, you're right.)

These other proteins include, first of all, an enzyme that makes a direct copy of the stretch of DNA that codes for a protein. This process is called *transcription* because the code is essentially the same, though the physical substrate carrying this copy is not DNA, but its cousin molecule, RNA (ribonucleic acid). This RNA copy, called a message, is then transported out of the nucleus of the cell—the information storage compartment—to the cytoplasm of the cell, which is where proteins are made. The RNA message is, like a tape, gradually passed through a complicated molecular machine called the ribosome (made of proteins and other types of RNA molecules) and at each successive three-letter code, the appropriate amino acid is added to the growing string that will form the new protein. This process is called *translation* because it takes the information in the language of nucleic acids and translates it into the language of amino acids. When the end of the message is reached, the protein is released, folds up into its predestined shape, and flits off to do its job, wherever in the cell it is needed.

But here's the rub—different cells need different proteins. Blood cells make hemoglobin but other cells don't. Immune cells make antibodies. Pancreas cells make insulin. Each different type of cell in the body—and there are many thousands of different types—expresses a different subset of the 20,000 proteins encoded in the genome. In fact, that profile of gene expression is precisely what makes muscle cells different from nerve cells or skin cells or blood cells.

So the DNA has to encode much more than just the recipes for each protein or active RNA—it also has to encode the instructions for when to make them, where to make them, how much of each one to make in any particular cell. This information is encoded in the sequence of DNA that flanks the part that encodes the protein itself. It is interpreted by other proteins in the cell, which seek out and bind to short stretches of DNA, promoting or inhibiting production of the RNA message from that gene. Each cell type makes a distinct set of these regulatory proteins that control and coordinate the expression of all the other genes in the cell. I should add that, in addition to protein-coding genes, there are several thousand other genes that encode RNA molecules that are not merely messengers, but that themselves have some active function in cells.

You may now start to see the central problem of developmental biology—how do all the cells "know" which genes they should turn on or off? If you start with a single cell—the fertilized egg—that divides over and over to make an embryo, how is it that cells on the outside turn into skin, while those on the inside turn into muscles or internal organs? How do we get the brain at one end and the tail at the other? (And, yes, you did once have a tail.) Another level of information must be encoded in the DNA—beyond turning single genes on or off (making the protein or not) in any given cell, this process must be coordinated for all the genes in each cell and for all the cells in the organism, in such a way that it can self-organize as it grows and as the various tissues differentiate.

So, we need to expand our molecular biology definition of a gene a little bit—it should be a physical stretch of DNA that encodes not just the amino acid sequence of a protein, but also the regulatory instructions of when and where that protein should be made (and likewise for genes encoding active RNA molecules).

Okay, at this point you may be asking: What has all this molecular biology got to do with the original definition of the gene, as a unit of heredity? And actually, so far, it has nothing to do with it, because we're still missing the most important element that links molecular biology to heredity—variation. If we return to Mendel's peas we can see this in action. Among the many traits Mendel studied was flower color—he had two strains of pea plants, one with purple and one with white flowers and his breeding experiments showed this was due to a single genetic difference. Over a century later, this genetic difference was finally identified and the biological mechanism underlying flower color elucidated. The purple flowers are purple because their cells produce a pigment called anthocyanin. To make this pigment requires the action of a suite of protein enzymes, each one encoded by a different gene. These genes are normally turned on in flower cells by the action of a regulatory protein. A mutation in the gene for that protein—a change to a single base of DNA, from a G to an A—stops the protein from being made, which in turn means the enzymes for anthocyanin formation are not expressed, and so the flowers remain white.

So, it is really that genetic difference or *variant*—the A version instead of the G version, in this case—that Mendel was studying. That is the crucial link between the concept of a gene as a unit of heredity and the molecular biology concept of a gene as a segment of DNA encoding a protein. And it is precisely those kinds of genetic variants that affect our traits. We all have a human genome, encoding those 20,000 proteins, but we don't all have exactly the same *versions* of each of those recipes.

VARIATION—THE KEY CONCEPT IN GENETICS

For example, we all have the gene that codes for hemoglobin, but some of us have a version with a different letter at a particular position in the DNA sequence, which causes a different amino acid to be inserted into the protein, which impairs its function, causing the disease sickle-cell anemia. So, from different perspectives, the gene "for" hemoglobin is also the gene "for" sickle-cell anemia. When we're talking about genes "for" traits or diseases being inherited, we are really talking about inheritance of a version that contains one of those differences in the DNA sequence.

So, where do these differences come from? Simply put, from mutation. Geneticists use the word "mutation" to refer to both the process whereby some change occurs in the DNA sequence and to the resultant change or difference itself. There are many sources of mutation. Thanks to comic books and movies, people often think of mutation as involving some external causative agent, like gamma rays or toxic chemicals. It is certainly true that such factors, or others like ultraviolet light, can indeed induce mutations, which is why they increase the risk of cancer. But it is also true that mutations just happen.

Whenever DNA is copied, when cells are dividing, some mistakes occur. The process is simply not 100% accurate. Our genome has three billion letters of DNA to be replicated—the enzymes that do that job are incredibly faithful but, still, some errors can arise each time a new copy is made. To put that number in context, the famously lengthy novel *War and Peace* has approximately 587,000 words. With an average of five to six letters per word, this amounts to about three million letters. Imagine if you had to copy *War and Peace*, by hand, letter by letter, but multiply the length by a thousand—that is the scale of the job that a dividing cell has to do when replicating its genome. You'd probably forgive yourself a few errors.

Most of the errors in DNA replication involve a simple change to one letter of the DNA code—perhaps an A is inserted in the new copy where a C should have been. Or sometimes a letter is left out or an extra one is inserted. These "point mutations" are fairly simple typos and the cell has proofreading enzymes and DNA repair enzymes that detect and correct many of these errors. But not all of them. A few creep through, just as I am sure some typos will creep through in this book. Figure 3.2 shows some of the different types of mutation that can occur.

There are also more drastic mutations that involve deletions or duplications of larger segments of DNA, affecting not just a single letter, but whole sections of chromosomes. These are more like missing or

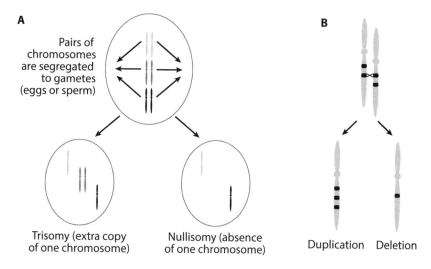

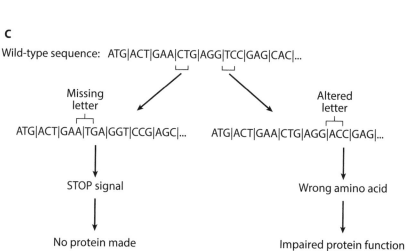

Figure 3.2 Types of mutations. **A**. Mis-segregation of chromosomes to gametes can lead to embryos with an extra copy or missing a copy of a chromosome. **B**. Misalignment of repeated sequences (black boxes) can lead to duplication or deletion of intervening segments. **C**. Errors in DNA replication can affect protein production or function.

duplicated pages in a book. They happen far less frequently than point mutations, but are harder for the cell to correct, and can have much more severe consequences.

Finally, the most large-scale disruption of the genetic information is when a cell inherits an entire additional chromosome or lacks an entire chromosome. Each of our cells contains two copies of each chromosome, one from our mother and one from our father. When cells divide in the body of a growing organism, each daughter cell also receives two copies of each chromosome. But when we make sperm or eggs, they get only one copy of each chromosome, so that, when they come together at fertilization, the resultant embryo has two copies of the entire genome again. However, sometimes when sperm or egg cells are being generated a mistake is made in segregating the chromosomes and either one is left out or an extra copy of one chromosome is put in. An embryo that results from one of these sperm or eggs would then have either only one copy or three copies of that chromosome instead of the normal two.

For most chromosomes in the cell that situation is not compatible with life. It simply causes too large a disruption to the cell's biochemistry. If a cell has only one copy of a chromosome, then it will typically make only 50% of the normal levels of each protein encoded on that chromosome. Conversely, if it has three copies, it will make 150% of the normal levels of the encoded proteins. Changing the levels of so many proteins at once, either up or down, drastically affects the balance of the various biochemical systems within the cell, to the extent that a viable embryo cannot develop. There are a few exceptions, however—embryos can survive with a change in copy number of some of the smaller chromosomes, though this does have deleterious effects. For example, Down syndrome is caused by an extra copy of the tiny chromosome 21. (The chromosomes are numbered in order of how long they are—chromosome 1 being the longest, and chromosome 22 the smallest.)

With all the things that can go wrong, you may be amazed that cells ever manage to replicate their DNA at all. Thankfully, millions of years of troubleshooting have led to the evolution of very robust systems to keep the occurrence of new mutations to a minimum. But they still happen, and when they do, we can ask: What happens to them? When a new mutation arises in an individual (because it occurred in the generation of the sperm or egg that produced that individual), what happens to it?

THE FATE OF NEW MUTATIONS

We are used to thinking about what happens to the individual in whom such a mutation is present, but it is equally important to consider things from the mutation's point of view, as it were, especially if we want to understand the origins and dynamics of genetic variation across the whole population. In fact, the answers to these questions—what happens to the individual and what happens to the mutation—are intimately related to each other.

In general, mutations could have a positive effect, no effect, or a negative effect on the organism's survival or fertility. Statistically, new mutations are most likely to be neutral, reflecting the fact that only about 3% of the genome actually comprises functional genes. That's the really important information—for most of the rest of the genome the particular sequence of the DNA bases doesn't really matter very much.

But in cases where mutations are not neutral, they are far more likely to be deleterious than advantageous. That is for two reasons. First, it is much easier to mess up a complicated system by random tinkering than to improve it. And second, it is because natural selection has already been doing this job for millions of years. It's hard to make a new mutation that natural selection hasn't seen before. And ones that increased fitness in a species would have tended to rapidly rise in frequency—so rapidly that they would often have become "fixed" in the population, outcompeting the previous version of that gene. It is that process of positive selection that leads to species divergence—the human genome is how it is because of the mutations that occurred in our ancestors that were selected for (along with a lot of mostly neutral mutations that came along for the ride).

At a molecular level, mutations can have an effect on genes in different ways. Let's consider so-called point mutations (changes to a single letter of DNA) first, as these are the most common kind. If these arise in the middle of a gene they may alter the sequence of the encoded protein, as with the example of hemoglobin above, where a point mutation leads to sickle-cell anemia. This can impair the function of the protein or even result in no functional protein being made at all. Mutations that occur in the regulatory regions of a gene can also have just as important effects, altering the DNA sequence that encodes the instructions for when and where to make the protein. On the other hand, some point mutations that occur within genes will cause no change to protein structure or expression levels and will be perfectly well tolerated.

The other major class of mutations is the larger deletions or duplications of sections of chromosomes. As with point mutations, these can affect genes or not, depending on which section of which chromosome is altered. And they can similarly affect protein-coding sequence or regulatory sequence (often both). Because of their size they are much more likely to affect some gene than any random point mutation is, and in fact often affect multiple genes that lie next to each other on a chromosome. For this reason, such deletions or duplications often have more severe effects. And it is the effect at the level of the organism that will determine the fate of the mutation.

If the mutation has a seriously deleterious effect—if it prevents normal development, or causes a severe disease early in life, or reduces the number of offspring of the person who carries it—then it will likely quickly disappear from the population. It simply won't be passed on to anyone if the person who has it does not survive or have offspring. From the population perspective, such a mutation will be a quick little flash that appears and disappears just as quickly. Figure 3.3A shows how selection impacts new mutations.

At the other end of the spectrum, if the mutation has no effect at all—if it doesn't affect the development or physiology or behavior or fertility or overall evolutionary "fitness" of the individual carrying it—then whether or not it gets passed on is simply a matter of chance. If that person happens to have children, some of them may inherit it. And if they in turn have children, then more people may inherit it. What will happen over a long time is that such a mutation may spread to more and more people, just based on the vagaries of genealogical successes in any given population—it will become a genetic *variant* in the population. If a given line dies out, so will the particular variants those individuals carried. If a line is more successful, evolutionarily speaking, then more of the population will end up carrying those variants that arose in that line.

For any given mutation, the likelihood that it will increase in frequency over time is thus directly related to what kind of effect it has on the organism. Very negative ones should remain very rare and often will

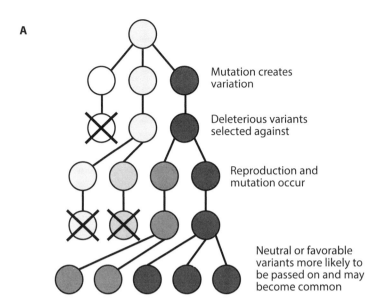

A

Mutation creates variation

Deleterious variants selected against

Reproduction and mutation occur

Neutral or favorable variants more likely to be passed on and may become common

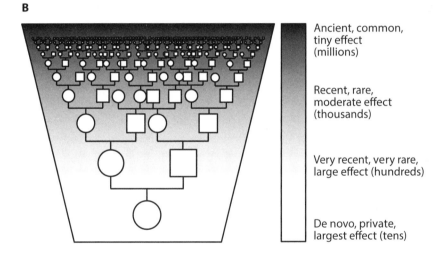

B

Ancient, common, tiny effect (millions)

Recent, rare, moderate effect (thousands)

Very recent, very rare, large effect (hundreds)

De novo, private, largest effect (tens)

Figure 3.3 The dynamics of mutation and selection. **A.** New variants enter the population in each generation due to mutation. Harmful variants are rapidly selected against. Favorable, neutral, or only mildly deleterious variants can persist and may become common. **B.** Any given individual (bottom circle) will carry a few new genetic variants, which may have large effects; some slightly older ones with smaller effects; and a vast number of common, ancient ones with minimal individual effects. (Panel A modified from Wikimedia Commons contributors, "File:Mutation and selection diagram NL.svg," *Wikimedia Commons, the free media repository*, January 30, 2017, https://commons.wikimedia.org/w/index.php?title=File: Mutation_and_selection_diagram_NL.svg&oldid=231578997; panel B modified from J. R. Lupski, J. W. Belmont, E. Boerwinkle, and R. A. Gibbs, "Clan Genomics and the Complex Architecture of Human Disease," *Cell* 147 (2011): 32–43.)

disappear in one or a few generations. Less negative ones may persist for longer and drift through the population, but still should not become very common. Ones that have hardly any effect on evolutionary fitness at all are free to drift to any frequency, just depending on chance, and may become common in the population.

OUR GENETIC HERITAGE

Now, let's flip our perspective. Instead of thinking about what happens to individual mutations going forward in time, let's look back in time and see what this means for each of us. How will this process, played out for millions of mutations across the whole genome in all of our ancestors, have shaped the genetic variation present in human populations right now? You can imagine that, if new mutations are introduced each generation, then a very large number of genetic variants must now exist in the human population. (I've switched to calling the mutations "variants," because some of them will have spread through the population so that we now have two versions at that site in the genome—the old one, and a newer one.) Of course, counteracting that trend is the fact that natural selection tends to weed out the variants that have deleterious effects. There is therefore a balance between mutation and selection that keeps the human genome generally intact, though at the evolutionary expense of many individuals.

So, in my genome, or in yours, we each have a very large number of genetic variants, many millions in fact. Many of these arose in a distant ancestor and subsequently persisted in the population for long enough that we eventually inherited them. If they lasted that long, they are also likely to have drifted to a higher frequency in the population generally. For example, at a given position in the DNA sequence along a given chromosome, 70% of the time there might be an A, while 30% of the time there could be a C. Sites like that are called single-nucleotide polymorphisms (or SNPs) and they are scattered across the sequence of the genome—about one in just over a thousand bases shows this kind of common polymorphism (meaning there is more than one version observed across people). That amounts to about 25 million such sites across the whole genome.

In addition to those kinds of SNPs, I may also have some variants that are quite different from the ones you carry—these may have arisen more recently, in the subpopulations and clans of my ancestors. Some of them may be relatively common in Europe, or even more specifically in Ireland, for example, but quite uncommon elsewhere in the world. You will have your own set of ancestral variants depending on your own ethnicity and ancestry. And, finally, each of us will have a set of rare or very rare variants—ones that arose very recently in our pedigrees, some of them even in the very egg or sperm cells that gave rise to us (see figure 3.3B).

If you compare any two (unrelated) humans, like you and me, you will therefore find millions of genetic differences between them—some common in the population and others rare. Because natural selection has had a long time to work on the common, ancient ones, we can safely infer that they probably do not have a large effect on any traits that affect fitness—not individually, at least. The rarer ones, though, are much less constrained in this regard. They can and often do have much larger effects, especially ones that have just arisen in the sperm or egg, as these have never been exposed to natural selection at all.

But what about their collective effects? If we return to the issue of explaining the variance in a trait across the whole population, what should we expect regarding the frequency of the genetic variants involved? Rare variants can have large effects but obviously each of them only has an effect in a small number of people. Common variants tend to have much smaller effects individually, but are obviously present in many more people, and since we all carry millions of them, their individual effects may combine to have a larger impact on a given trait.

INTERPRETING THE EFFECTS OF GENETIC VARIATION

So far, we have established that genetic differences contribute substantially to differences in psychological traits between humans—twin, family, and adoption studies prove that general point definitively. And we have begun to consider what those genetics differences actually are and where they come from. But that obviously leaves open the question of *how* they contribute to differences in our traits. By what mechanisms do the kinds of genetic variants discussed above lead to differences in our brains and minds? We will consider those mechanisms in detail for particular traits in the coming chapters, but there are a few general points to make here.

The most important point to bear in mind is that we are not talking here about how the human brain is built or how it functions, in a normative sense. We are merely talking about how it *varies*. So, if we find a genetic variant that causes a difference in, say, extraversion, that does not mean that we are somehow reducing human social interactions to the function of that single gene. A system can be highly complex—built from thousands of interacting parts and reliant on all of their functions, with emergent properties arising from their dynamic interactions and interactions of the system with the environment—and still be susceptible to variation in a single component. Those are not incompatible statements or conflicting viewpoints—they are complementary perspectives.

Consider a car. Modern cars are incredibly complex, with a huge amount of electronic control systems, in addition to the basic mechanical components. Any particular function that the car can perform involves some subset of these components, acting in a coordinated system. But if we remove or destroy a single component, any such system might fail or at least be impaired in its performance. If I took the spark plugs out of your car, you wouldn't be going anywhere. That doesn't mean that the spark plugs are the "component for going." Clearly, if you were just standing in your driveway holding some spark plugs in your hand, you wouldn't be going anywhere either. No one would claim that the locomotive capacities of a car derive solely from the spark plugs.

In the same way, finding a genetic variant that, say, lowers intelligence, does not imply that human intelligence can be reduced to the function of a single gene. No one would make that claim. No one is making that claim. Even finding many such variants in many different genes would not imply that human intelligence can be reduced to the functions of many genes. It would simply mean that variation in many genes contributes to the *variation* in human intelligence that is observed across a population. That is a much more modest, and indeed a much more precise, claim—one that is not in any way reductionist, as sometimes charged.

COMPLEXITY OF GENETIC EFFECTS

When we try to understand the mechanisms by which genetic variants contribute to differences in traits, we have to grapple with the complexity of the system. The relationship between specific genetic variants and specific traits is rarely as discrete as for the traits studied by Mendel. In fact, it is somewhat exceptional to find traits that are truly "Mendelian"—that are determined by two different versions of a single gene. Blood types are an obvious, but rare, example that follows Mendelian inheritance patterns. But even traits that we once thought were like that—things like eye color or hair color—have turned out to be more complicated. Importantly, they can be complicated in two quite different ways.

First, a trait can involve many different genetic variants *across the population*. So, having red hair, for example, might be caused by one genetic variant in one person and by another variant in another person. In each family, it might still be inherited as a Mendelian trait, but across the population it would look more complex.

Second, a trait may involve the effects of multiple genetic variants *in any individual*. This is by far a more typical situation. Height is a classic example. Mendel found a very unusual situation in his pea plants—rather than a continuum of heights, he had tall plants and short plants. When he crossed these together and intercrossed the offspring again, he still found either short or tall plants in the next generation. He didn't see the range of intermediate heights that you might expect. That is because there happened to be a single genetic variant in his population of plants that had a big effect on height, and he was therefore able to classify his plants into two clearly distinct bins.

This is not the typical situation in human populations. Instead, we see a continuous range of heights across the population, rather than discrete classes. If a tall person and a short person breed, their offspring will tend to fall somewhere in between their heights. Even if two tall people breed, their offspring will still show a range of heights, though they will tend to also be taller than average. This kind of "blended" inheritance is so ubiquitous for most traits in animals and plants, that it led to a serious difficulty for biologists in the early twentieth century in incorporating

Mendel's findings into a broader framework of how traits are determined. In particular, since most evolutionary change is believed to be very gradual, comprising tiny changes in phenotypes from one generation to the next, rather than sudden qualitative change, it was not clear that Mendelian inheritance had any relevance to evolution at all.

The answer to this problem is quite simple, but, like many simple ideas, it is only obvious in retrospect. It is that traits like height in humans are affected by the inheritance of multiple genetic variants at once. Each of these is still a discrete unit of heredity, as per Mendel's definition—it is just that their effects are not independent. Consider the situation where there are two versions of each of several different genes, one that tends to cause an increase in height (a "plus" version) and the other that tends to cause a decrease from the average (a "minus" version). Each of these may be inherited independently from the others. Even with only a handful of such variants in the population you will get a pretty smooth distribution of resultant phenotypes, reflecting how many "plus" relative to "minus" variants each individual inherits. Most people will be near the average and a smaller number will be near the extremes. Moreover, if genetic differences are only one of the sources of variance in the phenotype in question, then this continuum of outcomes will be smoothed out even more by variation in other factors.

However, this distinction between traits that show a clear qualitative difference (with two categories, like "tall" or "short") and traits that show a continuous distribution with only quantitative differences between individuals, is somewhat artificial and in fact is not mutually exclusive. Again, if we look at human height it is clear that most of the variation across the population is continuous. But it is also true that single mutations can have very large effects on height in individuals, producing clearly distinct outcomes of dwarfism or gigantism, conditions that are inherited in a Mendelian fashion. So both kinds of inheritance can be at play across the population at the same time.

HOW DOES GENETIC VARIATION INFLUENCE TRAITS?

Answering the question of *how* genetic variation leads to differences in traits is ultimately what the modern science of genetics is all about. In some cases, the link is pretty straightforward. For example, as mentioned above, a mutation in hemoglobin causes the disease sickle-cell anemia. The reason is pretty simple: the mutation alters the shape of the encoded hemoglobin protein, causing it to stick together in strands, in turn altering the shape of the red blood cells, which are impaired in doing their job of carrying oxygen around the body, resulting in the symptoms of anemia. The explanation for red hair is similarly straightforward: many cases of red hair are caused by mutations in a gene that encodes a protein called melanocortin 1 receptor, or MC1R. This protein acts in cells in the skin (including the scalp) to drive them to produce the dark pigment melanin. When it is mutated, these cells will produce a lighter, reddish pigment instead. In dwarfism, things are a bit more complicated, as it can be caused by mutations in any one of about 200 different genes. However, in most of these cases, the gene involved encodes a known growth factor or receptor, thus providing a direct explanation for reduced growth.

These examples are all fairly simple because the phenotype in question directly reflects the function of the mutated protein at the cellular level—whether it is carrying oxygen around the body, controlling pigment production, or encouraging skeletal growth. There are a few examples of behavioral traits where a similar situation applies, involving genes encoding proteins with quite specific cellular functions that directly regulate specific behaviors. For example, mutations in the gene encoding leptin are associated with morbid obesity, directly reflecting the function of leptin as a hormone that signals fat levels in the body and regulates appetite. Mutations in the *PER2* gene affect circadian rhythms and sleep patterns, because the PER2 protein is itself a component of the cellular clock system that keeps track of circadian rhythms. And there are many genetic variants affecting proteins that directly act as receptors for different kinds of sensory information; these mutations can lead to differences in our ability to smell or taste certain compounds, to feel cold or pain, or to distinguish different wavelengths of light and thus see the normal spectrum of colors.

But for most behavioral traits these kinds of direct links to the cellular functions of specific genes are not apparent. There is no obvious function at a cellular level that relates in this same way to traits like extraversion, or intelligence, or handedness. Nothing at a cellular level can explain the delusions and hallucinations and disordered thoughts that accompany psychosis. There are no genes for thinking straight, or not seeing things that don't exist, or not having strange beliefs. There aren't genes for language, or not obsessing about things, or not swearing uncontrollably. No proteins directly control what kinds of things you are interested in or your musical talent or how conscientious you are.

Those high-level functions and traits are, instead, emergent properties of complex neural circuitry within the brain—circuitry that is assembled by instructions from thousands of genes and which functionally involves the products of thousands of other genes. Variation in some of those genes can cause variation in how those circuits work, manifesting as differences in high-level mental properties, but it does so in a highly indirect manner.

In the second section of the book we will look at the genetics of diverse psychological traits and explore the underlying mechanisms through which the effects of genetic variation are manifested. In many cases, these effects are developmental. The program of development involves thousands of genes, interacting in extremely complex ways. It can therefore be affected by variation in all those genes, such that different individuals will have a different outcome—in the same way that the shapes of our faces are all unique, so are the structures of our brains. But, as for faces, the outcome of development is not completely determined by our genomes—it is simply constrained within a possible range. Inherent randomness in the processes of development themselves creates another important source of variation. In the next chapter we will consider how this variation impacts brain wiring and contributes to innate differences in psychological traits.

CHAPTER 4

YOU CAN'T BAKE THE SAME CAKE TWICE

～～～～～～～～～～～～～～

The evidence from twin and adoption studies shows that for most psychological traits and also for measures of brain anatomy or function, genetic differences make a major contribution to differences between people, while differences in our family environments play a minor role, if any. But that is not the end of the story. Those two sources of variance do not fully explain all the variance in the traits across the population, even when you combine them. Something else is having an effect. Despite having both identical genomes and a shared family environment, MZ twins do not have identical values for psychological traits. For some traits the correlations are not even that high—they are much higher than for DZ twins or other siblings, but are often on the order of 0.4 to 0.5. Clearly, we're missing something important. There must be some other factor that tends to make MZ twins different from each other and that must also therefore be an important source of variance across the population generally.

THE MISSING THIRD COMPONENT OF VARIANCE

Regrettably, this factor is referred to in the behavioral genetics literature as the "nonshared environment." I say regrettably because that term connotes an effect that comes from outside the person—that is, in the environment. It suggests that whatever that factor is, it should contribute to the "nurture" side of the famous phrase, rather than the "nature" side. Indeed, it has been claimed that the findings from behavioral genetics actually offer the strongest evidence for environmental effects on psychological traits. Because these traits are not completely heritable, the conclusion has often been that the remainder of the variance must be due to environmental factors. In fact, this assumption is not justified. There are many possible factors that could contribute to this unexplained variance, and little evidence that it should be thought of as truly environmental.

The first is that the tests and measures we are using in the psychological domain may simply not be very accurate. That itself is something we can measure, by having a person take the same test multiple times and seeing how consistent the results are. While it certainly makes a contribution, the variability in test results for individuals is not sufficient to explain all the additional variance observed. Test-retest reliability (the correlation between test results for an individual person who takes a test twice) is around 0.9 for IQ tests, and typically on the order of 0.7 for things like personality trait measures. (A correlation of 1.0 would indicate perfect agreement between the two test sessions.) Those numbers obviously place an upper bound on how similar we can expect measures to be between MZ twins—if the same person tested twice only shows a correlation of, say, 0.8, then we couldn't expect MZ twins to show any higher correlation. These results suggest that typically between 10% and 20% of the observed variance in psychological traits may be attributable to measurement error.

The second interpretation of the effect of the "nonshared environment" is that it does, in fact, reflect environmental or experiential factors, but ones that are unique to individuals, rather than those that are shared due to a common family environment. On the face of it, this sounds kind of plausible. But if you dig deeper, it really starts to seem contrived. It suggests that my psychological traits can be affected by my experiences, but only if my cotwin does not also have those experiences. Because if we were similarly affected, then that would make us more similar to each other and more different from everyone else—that is, it would show up in the shared family environment term. It also seems to say that some environmental factor can cause a difference between two people if there is variance in exposure to that factor within families but not if that variance occurs between families—again, the latter effect would show up in the shared family environment term.

The idea goes that interactions with peers, teachers, or other kinds of experiences outside the home can have a much larger effect than interactions within the home. But if being reared in different families has so little effect on our psychological traits, then why should we think that interactions with peers could have such a big effect? Proponents of this interpretation suggest that our experiences are so unique that they can only make us different from other people—they cannot make us more similar. But these are two sides of the same coin. If there is anything systematic at all in how nurture or culture exert their supposed effects, this should be manifest in an increased similarity of people who grow up together. We may not have exactly the same experiences but we can certainly have similar *types* of experiences. After all, siblings reared in the same family are also more likely to share peers, schools, and a wider community and culture, not just a shared home environment.

Another proposal is that there may be systematic differences within families that make children in the same family less similar to each other. But if we consider, for example, the idea that differential parenting might have such an effect, then why would it not make children in different families even less similar to each other? Surely their parenting must be even more different if they actually have different parents. The only way to rescue this idea is to propose that is not the way your parents treat you that matters, it is *the fact that* it is different from how they treat your siblings that has an effect. Under that model, if you didn't have any siblings, there would be no opportunity for such effects to arise. Again, this idea seems hard to take seriously.

We do not have to rely on argument here as there is no shortage of empirical data. Many studies have looked for systematic associations between specific environmental factors or experiences that differ between siblings and specific behavioral outcomes. These typically fall under a number of categories including differential parenting, peer relationships, sibling interaction, teacher relationships, and what is known as "family constellation" (birth order, age difference between siblings, whether or not they are the same gender, etc.). The results from these studies are very clear. They have failed to identify any robust, consistent, or substantial effects on any of a variety of outcomes including adjustment, personality measures, or cognitive ability.

Overall, there is thus very little reason to think that that these kinds of nonshared experiences have a large effect on our psychological traits and no direct evidence to support the notion that they can explain the residual variation between MZ twins who are reared together.

The final possible factor contributing to this unexplained variance is something else entirely. It is not, in fact, "environmental" at all, but intrinsic to each person, arising from inherent randomness in the processes of brain development. While your genotype encodes the program to build a human being like you, it does not encode the instructions to build you specifically. If we started again and let the embryo that gave rise to you develop again, "you" would not be the result (not even "baby you" would be the result). Your clone would be the result, but he or she would be different from you in many ways.

The complex machinery of the brain emerges from instructions encoded in the genome, but it is not mapped out there like a blueprint—there is no one part of the genome that corresponds to one part of the brain or one type of nerve cell. It is more like a recipe, or a series of protocols, which, when carried out faithfully, result in a human being with a human brain. And, just like a recipe, no matter how detailed and precise it is, there will inevitably be some differences in the outcome from run to run—you can't bake the same cake twice.

Across the population, developmental variation may explain a large amount of the "nonshared environment" variance in phenotypic traits that is not due to either shared genes or shared family environment or even to unique experiences. This variation is thus not extrinsic to the organism at all and should be considered not on the "nurture" side but on the "nature" side of the ledger, in that it contributes to innate differences between people.

To understand where this variation comes from and why it can have such large effects, we must consider the processes and machinery of neural development—what are the physical mechanisms involved in generating the incredibly complex cellular architecture of the human brain?

GROWING A HUMAN

We'll have to start at the very beginning, with a single cell, a fertilized egg, that contains within it a human genome, with all the instructions to build a human being. Those instructions are made up of all the genes that code for proteins or RNA molecules—the little molecular machines that will do all the work in the developing embryo. But the instructions also crucially include the regulatory sequences of DNA that specify where and when to make each protein. Indeed, if we want to look for the crucial differences that specify making a human brain instead of a chimp brain, most of them will be in those regulatory sequences. The proteins themselves tend not to vary that much—in evolutionary parlance, they are highly conserved. So much so that, in many cases, they can often be experimentally substituted from one species to another and still work fine—even across species as diverse as mice and flies, for example. What tends to differ far more is the precise control of how all those proteins are expressed.

Now, we must keep in mind that our brand new egg doesn't just have some generic human genome—it has its own brand new genome, carrying not just all those universal instructions but also a unique combination of genetic variants that has never been seen before and will never be seen again. Those variants can affect the sequence of proteins, changing how they work, or can affect the regulatory elements, altering the precise patterns of gene expression. All of those differences can affect the final outcome of development and are the source of the heritable differences in brain structure that we have been considering.

Our single cell rapidly divides to make two, then four, then eight, and so on until we have a little round ball of maybe a thousand cells. From the outside, these still all look the same, but there is already a lot happening inside the embryo that is laying down the eventual pattern of the organism—in this case, a human baby. Cells are already starting to differentiate from each other, depending on where they are in the embryo—ones on the outside will form skin and nervous system, while others that migrate into the middle of the embryo will form muscles, bone, and blood, and a third layer will make the internal organs. The axis from the head to the tail is also already specified, as is the axis from the back to the belly. That patterning derives from small initial differences between cells in the very early embryo, in mammals partly depending on a sort of molecular memory of the precise point where the tiny sperm entered the relatively gigantic egg.

Even as the fertilized egg makes its first division, those two resultant cells are already different from each other—they are already expressing a different profile of genes, making more or less of each of the 20,000 proteins encoded in the genome. The job of many of those proteins is to regulate the expression of other proteins. As a result, small initial differences in a couple of genes can rapidly be amplified through a network of complex feedback interactions to result in quite different overall profiles of gene expression between two cells. That, in a nutshell, is how cellular differentiation occurs—muscle cells express a different profile from skin cells or liver cells. The key to development is organizing those cells spatially, so they are all in the right places. For that to happen, the cells have to talk to each other—they need some information about where they are in the embryo and what kind of cell they should turn into.

That is accomplished through the actions of proteins that are made in cells at one point of the embryo, but that are pumped out of those cells, diffusing through the embryo such that the concentration is very high near where they are made and then declines smoothly farther away in the embryo. The concentration of these proteins is detected by receptors on cells, and that signal is transmitted internally to control the particular profile of genes turned on in any one cell. In that way, we get brains and hearts and limbs and eyes all made in the right places.

The amazing thing about this process is that none of the individual cells knows the plan. It all just happens through a series of mindless biochemical interactions, with each cell reacting to signals from outside it, turning on some genes and turning off others, and then passing that information on to its descendants as it divides and the embryo grows. Each cell carries all the information to make the whole organism but none of them sees it. They're like actors in a massive ensemble cast, all of whom know their own lines and their cues, but none of whom sees the whole script. In fact, the only one who gets to see the final production is the ultimate critic: natural selection. Good scripts survive; bad scripts die (or close after a limited run).

PATTERNING THE BRAIN

The processes involved in patterning and differentiation in the early embryo are reiterated as organs begin to form and themselves become patterned. Nowhere is this more evident than in the brain, which has vastly more subregions than any other organ. This process happens by successive subdivision—first, the forebrain is set apart from the midbrain, hindbrain, and spinal cord. Then new patterning centers at the borders between these regions produce new signaling molecules that allow further subdivisions of each of these fields, and so on and so on, eventually producing all the different structures of the brain, organized in the right way. Each subregion has its own distinct cell types, usually of many different kinds. Even the retina, for example, which is just one small part of the central nervous system, has at least 200 distinct cell types that we know of.

A common misconception about brains, often portrayed in artists' renditions or animations of what is going on in there, is that all neurons are the same and they are laid out in a random fashion, in what amounts to a kind of a spongelike structure, with each neuron simply connecting to its nearest neighbors. Nothing could be farther from the truth.

There are in fact hundreds of different types of neurons and many thousands of subtypes. Each of these has its own specific morphology, biochemistry, electrical properties, and patterns of connectivity with other cell types. Neurons are polarized information-processing devices—they have an "in" end and an "out" end. In the middle sits the cell body, which contains the nucleus of the cell, where the DNA is, and a lot of the general metabolic machinery that every cell has. It is the long cellular fibers that extend out from a neuron that make them special. From one end, the neuron extends a tree of dendrites (literally, "branches"), which collect information—they are the input side. From the other end, it extends a single fiber, called an "axon," which is the output side of the neuron. It may project only locally or extend over very long distances and can branch to form connections with many other neurons, or, in some cases, with muscles. The shapes of different types of neurons are extremely diverse, with variation in how big the cell body is; how extensive the dendritic arbor is and what shape it takes (bushy, fernlike, treelike, etc.); whether the axon is long or short, thick or thin; and many other properties (figure 4.1).

Neurons also differ tremendously at the biochemical level, in ways that determine how they process information. Neurons conduct signals by electrical currents, mediated by the flow of charged particles (mainly sodium, potassium, calcium, and chloride ions) in and out of the cell, through regulated pores called ion channels. These electrical signals

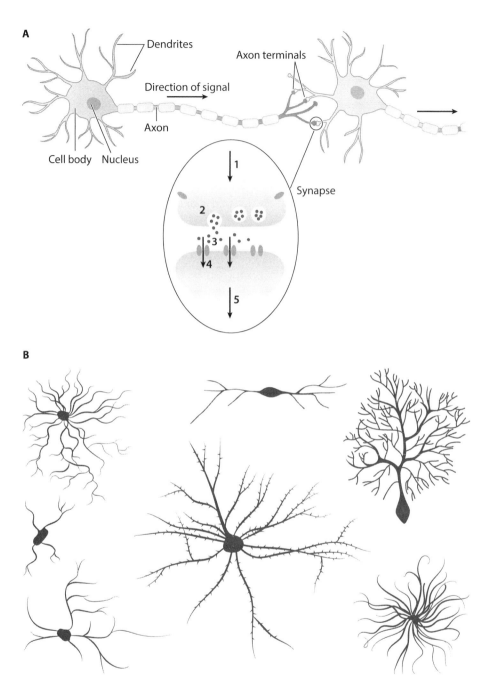

Figure 4.1 Neurons. **A**. Neurons collect information through their dendrites and transmit it down their axon to other cells, via synapses. At the synapse, an electrical signal (*1*) triggers release of neurotransmitter molecules (*2*) that are detected by receptor proteins on the postsynaptic side (*3*) leading to influx of electrically charged ions (*4*), which can trigger a new electrical signal (*5*). **B**. A small sample of the astonishing diversity of neuronal morphologies.

typically cannot pass directly from one neuron to another, however, as each cell is enclosed in its own cell membrane and separated from other neurons. Special structures called *synapses* have therefore evolved as sites of communication between them. At a synapse, an electrical signal being conducted down the axon is converted into a biochemical signal—if there is enough electrical current, the synapse will release a small packet of molecules called neurotransmitters. These neurotransmitters are detected by special receptor proteins on the other side of the synapse—that is, on a dendrite of the next neuron. If enough neurotransmitter is detected, then the next neuron will initiate an electrical signal of its own.

Each type of neuron is characterized by a specific profile of ion channels, neurotransmitter receptor proteins, synaptic plasticity proteins, and many other types of protein that collectively determine the electrophysiological properties of the cell. Some neurons require a strong incoming signal to become activated themselves, while others are far more sensitive. And some become more sensitive with repeated stimulation, while others turn down their responsiveness. Perhaps the biggest difference lies in the neurotransmitter that each neuron releases. I talked above about neurotransmitter release as if it always tends to make the downstream neuron fire an electrical signal. But actually some neurotransmitters do exactly the opposite—when they are released they tend to inhibit the downstream neuron from firing. If you think about it, you can see why this kind of neuronal signal is so crucial—if neurons only ever excited each other, then whenever one neuron became activated the signal would simply spread through the whole brain like a wildfire until they were all active and we would constantly be in a state of epileptic seizure. Instead, at any given moment, each neuron is integrating the level of excitatory and inhibitory inputs, and the balance between these determines whether or not it will fire a signal.

All these different types of neurons are laid out and interconnected with exquisite specificity. Each region of the brain has its own cellular architecture, with specialized excitatory and inhibitory neurons arranged in specific ways, creating local microcircuits designed to process particular types of information and carry out specific types of computations (see figure 4.2). The circuits that process visual information, for

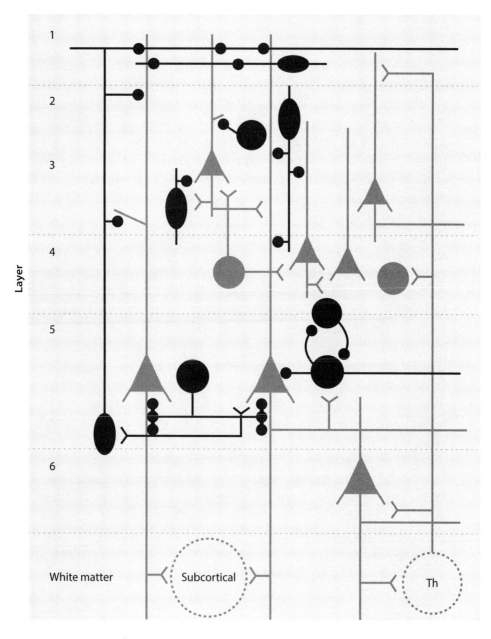

Figure 4.2 Neuronal circuitry in the cerebral cortex. Dozens of different types of excitatory (*gray*) and inhibitory (*black*) neurons are interconnected to carry out particular computations on information coming in from other areas of cortex, thalamus (Th), or other subcortical regions. (Modified from Z. J. Huang, "Toward a Genetic Dissection of Cortical Circuits in the Mouse," *Neuron* 83, no. 6 (2014): 1284–1302.)

example, are wired in a different way from those that process smells or sounds, because the incoming information, the salient features of it, and the types of computations that must be performed on it are all radically different in each case.

So, the challenge for the developing brain is, first of all, to make all those different types of cells, and, second, to get them organized in the right ways in each brain region. This is made even more challenging by the fact that many cells are not born exactly where they are needed. In most cases they must migrate some distance from where they are born to take up their appropriate final positions in the cellular architecture of their target destination. For example, in the cerebral cortex, different types of cells migrate outward in a distinct sequence to form a six-layered structure—each layer comprises different cell types, with different jobs to do in the final circuitry. These are just the excitatory neurons, however—the inhibitory neurons of the cerebral cortex are born in a completely different part of the brain and must make a much longer journey to reach the cortex and eventually become integrated into cortical circuits.

If you've ever watched simple, single-celled creatures like amoebae or bacteria under the microscope, you may have seen that they can move around in response to cues in the environment. They can swim or crawl toward a food source, for example, or away from noxious chemicals. Migrating cells in the developing brain do exactly the same thing—the only difference is that their environment is made up of other cells in the brain and the soup of proteins that they produce. Some of those proteins act as signals, attracting or repelling migrating cells, depending on the receptor proteins that each cell expresses. As with patterning of organs and tissues, these mindless biochemical interactions accomplish a remarkable feat—the ultimate emergence of a stereotyped and complex structure.

Okay, so now our growing embryo is starting to look pretty good—it's got its head at one end, its tail at the other, and all its limbs and organs in the right places. And its brain is nicely organized too, with its hippocampus over here and its cerebellum over there, and all the different cell types laid out just so. But that's only half the battle. Now all those bits and all those cells need to get connected to each other in the right ways, and that requires a whole other set of instructions.

WIRING ITSELF UP

Once a nerve cell has been born and migrated, like a little amoeba, to its correct position, it settles down and starts sending out the cellular protrusions that will form its dendrites and its axon. These don't just emerge at random—they are highly specified from the start, with dendrites forming in a particular pattern, depending on the cell type, and with the axon extending along a highly stereotyped pathway. Each growing axon is tipped by a remarkable structure called a *growth cone*—a lively little thing that sends out feelers exploring its environment and that has its own little motor, pulling the growing axon behind it. These growth cones are guided by signals much as migrating cells are—proteins that are secreted from or displayed on the surface of other cells. Each growth cone, in turn, expresses a distinct set of receptor proteins on its surface, which determines its individual responses to these cues in its environment.

You can imagine the scene, as literally billions of nerve cells extend their axons, with their growth cones all shmooing around at the same time, looking for their targets. The process is not so chaotic, however—it's quite organized, in fact, initially by the same kinds of diffusible proteins that confer some pattern on the early embryo. Each growth cone can orient to gradients in the concentration of proteins coming from the front or the back of the brain, from the top or the bottom, from the middle or the edges. And very rapidly large highways form—tracts of nerve fibers that individual axons can join and follow and then exit from at an appropriate point, just like getting off a highway (see figure 4.3).

Now that we've got the growth cones in the right neighborhood, we need to direct them to the right partners to make synaptic connections with. Again, this process is incredibly highly specified—neurons don't make connections with just any other cell; they are very selective in their choice of partners. Not at the level of individual cells, at least not in complex nervous systems like ours, but certainly at the level of *cell types*. By now you should be able to guess how that specificity is accomplished— each cell type expresses a different set of proteins on its surface, which are detected by receptor proteins on the surface of growth cones. The particular profile of cues and receptors will determine whether a synapse is made between any two cells.

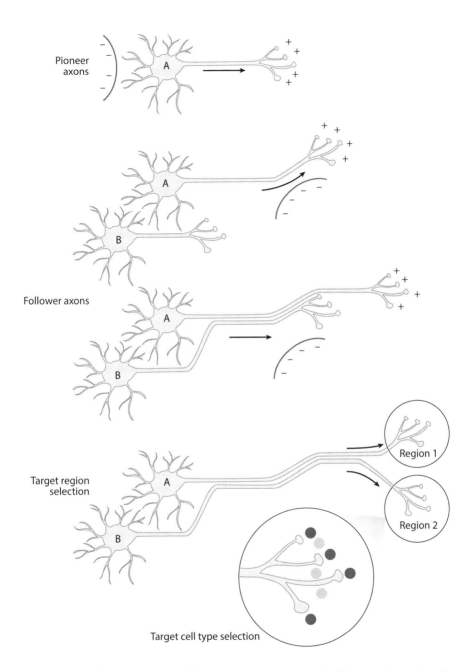

Figure 4.3 Guiding growing nerve fibers. Growing axons are guided by attractive and repulsive signals, which they detect with specialized receptor proteins. Different neurons express distinct repertoires of receptors and therefore follow different trajectories and select different target regions and target cell types.

These, then, are the processes by which the circuitry of the brain assembles itself. They are complex, but not magical. They come down to individual biochemical mechanisms that can be directly observed and studied in cells. You can grow neurons in a dish and directly see that different types of cells or their axons are attracted to or repelled from certain proteins, or that exposure to one or another particular protein will make them stop and make a synapse. It is the way that these processes are coordinated that seems almost miraculous, but that's what billions of years of evolution will get you—a developmental program encoded in the genome that is capable of directing the self-assembly of a structure of ironically mind-boggling complexity and precision.

INDIVIDUAL DIFFERENCES IN BRAIN DEVELOPMENT

The interesting thing, of course, is that your program is different from mine. Some subset of the millions of genetic differences between us will affect the genes that carry out these developmental processes. These variants may alter the amino acid sequence of the encoded proteins or change when and where they are expressed, in turn affecting the outcome of brain development. In fact, it is precisely by studying the effects of such mutations in animals like flies and mice that developmental neurobiologists have worked out the principles of brain development described above and identified many of the specific genes involved. The ones that we know the most about in humans are the ones that have the most dramatic effects, and not in a good way.

There are hundreds of known genetic disorders that affect the processes of brain development, resulting in brain malformations with concomitant neurological, psychiatric, and psychological effects. These may arise from mutations in genes involved in brain patterning, control of proliferation, migration of neural cells, guidance of growing axons, or many other developmental processes. Such mutations can lead to malformations that are even evident on MRI scans—undergrowth or mispatterning of specific brain areas, clumps of misplaced cells in the wrong places, or failure of specific nerve tracts to form, such as the connections between the cerebral hemispheres or the tract from the cerebral cortex down the spinal cord.

Other effects may only be evident if you look under the microscope at sections of the brain following surgery or postmortem, such as altered layering in the cerebral cortex or subtle disorganization of the cellular architecture of other structures. Mutations in genes controlling the formation of synapses may not be evident even in that kind of histology, but can have equally severe effects on the function of specific neural circuits and brain systems. And there are many additional disorders caused by mutations in genes with much more generic functions—like metabolic enzymes, for example—which are not directly involved in neurodevelopmental processes but are nevertheless required for them to be carried out properly. We will consider these kinds of serious neurodevelopmental disorders in more detail in chapter 10.

Thankfully, though, most of the genetic variants that we all carry will have much more subtle effects on brain development, contributing to variation across the normal range, rather than causing overt pathology. The combined effects of many such variants are responsible for differences in overall brain size, in the relative sizes of various brain regions, in the amount or organization of nerve connections between areas, or any of the other physical traits of brain structure that we have seen are highly heritable.

However, differences between the developmental programs encoded in each of our genomes are only the start of the story of what makes each of our brains so unique. The particular run of the program that gave rise to each of us involved a series of events that could never and will never be repeated. This leads to variation in outcome that we can see directly in the brains of MZ twins. These are strikingly similar to each other, both in physical structure and functional organization. But they are not completely identical—even at birth, the structures of their brains already show some differences, just as the structures of their bodies and faces do.

We don't even need twins to see the effects of this kind of variation—they are also visible within individuals, if we compare the two sides of our bodies. The two sides of the body develop from largely independent runs of the developmental program encoded in each of our genomes. This results in minor variation from side to side in things like how long your arms or legs or fingers or toes are, or whether one foot is slightly bigger than the other, as well as variation in the precise layout of things like blood vessels or the hairs on your hands.

Figure 4.4 Facial asymmetry. The center picture of former US President Barack Obama can be split into two halves. Mirror images of the left and right sides highlight the differences between left and right. Center panel from Wikimedia Commons contributors, "File:Barack Obama.jpg," Wikimedia Commons, the free media repository, Feb 17, 2016, https://commons.wikimedia.org/w/index.php?title=File:Barack_Obama.jpg&oldid=187747492

This variation is most striking in our faces. You may not appreciate it, but your face is probably quite asymmetric. It is difficult to notice this when we look at a whole face because our brains are practiced at processing the entire picture—the "gestalt"—of an individual's face. But there is a simple trick to illustrate it. If you take a selfie straight-on and bisect the image down the middle of your face, you can split it into a right-side half-face image and a left-side half-face image. If you then make a mirror image of each of those and put the two mirror images back together, you can see what you would look like if you were a symmetric version of your right-side or of your left-side self. The result is usually quite striking—these two faces often look like clearly different, yet eerily similar people (see figure 4.4).

The fact that we see variation across the two sides of the same individual illustrates a really crucial point: this kind of variation is not coming from some outside factor. It is not "environmental" in origin, as the terminology used in twin studies to refer to any nongenetic variance would suggest. It is *intrinsic* to the developing organism itself. Consideration of the complex processes of development outlined above gives insight into the sources of this variation.

NOISY CELLS AND NOISY GENES

Because the processes of development operate at a molecular level, they are susceptible to what engineers call "noise" in the system. This means there is variation from moment to moment in the precise numbers and positions and states of all of the millions of individual protein molecules and other cellular components that carry out the neurodevelopmental processes described above.

The information in the genome can specify approximately how much of each of the 20,000 proteins to make in each cell and there is a sophisticated protein-trafficking machinery that gets each of these protein molecules to the part of the cell where it is needed. But after that the genome has no control over the precise location or biochemical state of every molecule. They will buzz and jitter and jiggle around essentially at random, bouncing into each other with a certain probability, binding to each other with a certain affinity, and catalyzing chemical reactions with a certain rate.

Now, it's possible that all of these movements and interactions and reactions are really deterministic—that if we knew the locations and states of every molecule at a given moment, we could predict their future states with perfect accuracy. When I say that there is intrinsic randomness in the system, perhaps that is just an expression of ignorance of the true sources of variability. Whether a system like this displays true randomness, or even whether true randomness exists in the universe at all, is still debated by physicists.

Perhaps all this noise has its ultimate origins in fundamental indeterminacy at the most basic quantum level. This means that the exact state of the system cannot be defined at any given instant—in fact, some would argue it is not right to think of the system as even having an exact state at any given instant, not at the finest level of detail at least. Or perhaps it arises at a higher level, in the way that molecules tend to jitter around, in effectively random thermal fluctuations (called that because the jitter increases at higher temperatures, like our body temperature). The diffusion of molecules within and between cells is also subject to a high degree of what is arguably real randomness.

As it happens, the answer to this debate really doesn't matter. Whether the system contains true randomness in the metaphysical sense or is just so complex that its exact state at any given moment is unpredictable, the result is the same. The point is that *the genome* cannot predict (and cannot specify) any cell's exact state—certainly not every cell's exact state. There is noise in all the operating parameters of every cell—in the concentration of every component, in the flux of every reaction, in the state of every control system.

Now, you might think all that might not matter very much at the level of a whole cell, but actually that noise can percolate through the whole system and manifest in big fluctuations at the level of things that cells really do care about, like gene expression. As described previously, the first step in converting the DNA code into protein molecules is to transcribe a copy of that code to produce a messenger RNA (mRNA) molecule. That act of transcription requires the binding of many different regulatory proteins and enzymes to the DNA, which entails myriad atomic interactions and biochemical processes that ultimately regulate the amount of mRNA made from each gene.

These processes are all subject to noise at the molecular level, the effects of which are evident if we look at how mRNA molecules are actually made. In any given cell, some genes will be expressed at a higher level than others. Over a longish time frame, we will just see an apparently steady rate of production of mRNA from any particular gene. But if we look in individual cells over much shorter time frames we see something very different—mRNA molecules are transcribed in bursts, with periods of quiescence in between. When a gene is turned on, by the actions of regulatory proteins for example, what is really happening is that the overall frequency of bursting is increased (i.e., it spends more time bursting and less time quiet). But at any given moment, whether or not the gene is bursting is probabilistic—that is, it has a large essentially random component to it. We can see this directly in single cells that have two copies of any given gene, one on each chromosome. While the overall frequency of bursting may be equivalent across the two copies, the precise pattern of bursting is largely independent. The actions of regulatory proteins increase or decrease that probability, but the remaining random noise means that the precise numbers of mRNA molecules produced can fluctuate considerably over time.

Such fluctuations in gene expression can have surprisingly large effects, due to the complex network of positive and negative feedback interactions between different genes. If expression from each gene just fluctuated independently of all the others, this noise probably wouldn't matter much. But they are not independent. Quite the opposite in fact—they are extremely interdependent. If, by chance, amounts of mRNA and protein molecules produced from gene A reach a certain level, this may tend to increase expression of genes B and C, while decreasing expression of genes D and E, which can each have knock-on effects on other genes, and so on. These cross-regulatory interactions can amplify small initial differences in specific genes, leading to quite different global profiles of gene expression. These differing states can even persist for long enough to affect how a given cell will differentiate or what kind of progeny it will produce.

THE PROBABILISTIC NATURE OF NEURAL DEVELOPMENT

All of the processes of neural development described above—patterning, proliferation, differentiation, cell migration, axon guidance, synapse formation—rely on differential gene expression and on interactions between proteins (signals and their receptors, to begin with, as well as all the internal pathways of proteins that mediate the reactions to such signals). This means that each of these processes is subject to noise at multiple levels. As a result, none of these processes in the developing embryo is deterministic.

The probabilistic nature of neurodevelopmental processes becomes most obvious in the presence of mutations that slightly impair them. When the genes controlling processes like the migration of neurons or the guidance of their axons are mutated, what typically happens is that some of the cells or axons are misplaced, but some of them can still make it to their correct destination, even if some of the information they normally rely on is missing. The outcome for each individual cell or axon is probabilistic—the genetic variants determine the probability, but the actual outcome in a given cell is affected by noise in thousands of biochemical parameters that constitute its internal and external environment during this process. As these probabilistic events play out differently across individuals, some can end up showing a more severe overall phenotype than others, just by chance. If, for example, the average number of misplaced cells in some brain region is, say, 30%, some individuals may show 20% and others 40%.

Exactly how this process plays out can have serious consequences. There are, for example, a number of clinical conditions that affect migration of neurons in the cerebral cortex, resulting in large-scale cortical malformations or in smaller clumps of misplaced cells. The severity of these effects and the concomitant effects on intellectual functioning or neurological and psychiatric symptoms vary considerably across different patients. Of course a lot of that variation is genetic in origin—due to the background of other genetic variants each patient carries. But large variation both in anatomy and clinical symptoms is also sometimes observed even between MZ twins, reflecting the probabilistic relationship between the starting genotype and the final phenotype.

In addition, as these processes of cell migration play out independently across the brain, different brain regions may be more or less affected in any individual. For disorders causing clumps of misplaced cells, the distribution of these clumps is thus quite random. This too can have important effects, as such clumps of cells can disrupt electrical signaling in the brain and cause epileptic seizures. Indeed, if we look at the heritability of epilepsy, we see that if one MZ twin has the condition, there is a 30%–40% chance that other one will too—much higher than for DZ twins, showing the strong genetic predisposition for the disorder. But the precise location of the epileptic focus in the brain (whether in the frontal, temporal, parietal, or occipital lobe) is much less heritable— hardly at all in fact, reflecting instead the random expression of that risk across the developing brain.

Even in the absence of serious mutations, the generally probabilistic nature of the processes of brain development will contribute to quantitative variability in the precise numbers of neurons or axons or synapses in a given brain region. These parameters underlie the kinds of macroscopic structural variation that we can measure with brain imaging— differences in sizes of brain regions or connections between them. This means that intrinsic developmental variation is the most likely source of nongenetic variation in these parameters, which, as we have seen, can be quite substantial.

THE GARDEN OF FORKING PATHS

Variation in the neuroanatomical traits described above is continuous in nature, generating a smooth range of values across individuals. However, due to the self-organizing and contingent nature of the processes of brain development, this kind of noise can lead not just to a little quantitative fuzziness in brain structures, but also to qualitatively distinct outcomes. Formation of the connections between the cerebral hemispheres illustrates this point. The two halves of the cerebral cortex are connected to each other by a thick band of nerve fibers known as the corpus callosum (or "tough body," and you can imagine how that was found out). In humans this contains around 250 million axons, each projecting from the left to the right side, or vice versa.

You might expect that the kind of developmental noise that I've been talking about couldn't affect this structure dramatically because it should average out across so many axons, but this is not the case. The reason is that the formation of this structure relies on a number of prior events that involve a very small number of cells. Prior to any axons crossing the midline, the two hemispheres initially get connected by a small population of nonneuronal cells that lie adjacent to the midline on each side. This creates a small cellular bridge that the first axons use to cross what is otherwise an impassable gap. The axons that pioneer this route then act as a scaffold for the millions of axons that follow.

If that cellular bridge does not form, then, in many cases, no axons will cross the midline (see figure 4.5). Axons that reach that point will either whorl back on themselves, rather ineffectually, or sometimes will divert to an alternate route, much lower in the brain—a more ancient pathway that is a holdover from our evolutionary past. The contingent nature of this process, this reliance on a previous event having happened correctly, means that formation of the entire corpus callosum is sensitive to noise that affects only a small number of cells at an earlier stage of development.

Under normal circumstances, this cellular bridge forms almost without fail. But there are some mutations that can cause the process to fail, resulting in agenesis, or lack of formation, of the corpus callosum. This doesn't always happen though, as studies in mice have shown. Some

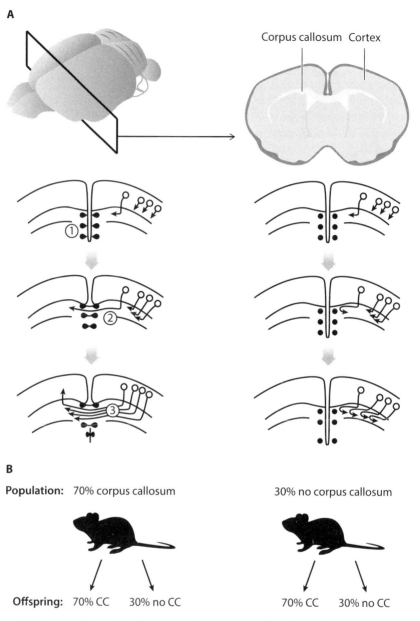

Figure 4.5 Corpus callosum development. **A**. *Top*: a section through an adult mouse brain shows the cerebral cortex covering the two hemispheres and the corpus callosum connecting them. *Bottom left*: the stages of normal corpus callosum development. Midline cells fuse and form a bridge between the two hemispheres (*1*). Pioneer axons cross (*2*). Follower axons cross (*3*). *Bottom right*: when the midline cells fail to fuse, pioneer and follower axons fail to cross, resulting in absence of the corpus callosum. **B**. In some mouse strains a proportion of animals end up with no corpus callosum (CC), despite every animal being genetically identical. This probabilistic effect is inherited regardless of the phenotype of the parent.

lines of mice carry such mutations and even though every animal in the line is genetically identical (as they've been bred that way over hundreds of generations), some of the individuals develop a quite normal corpus callosum, while others develop none. This isn't due to some environmental factor such as a difference in conditions in the womb, as it is observed across animals from the same litter. And it isn't due to the presence of other mutations, because if you breed individuals either with or without a corpus callosum, their offspring still show the same bimodal distribution of phenotypes (some with, some without), regardless of the phenotype of their parents.

It is thus the expression of a probabilistic event—the bridge either forms or it doesn't, with a certain probability. In "wild-type" mice, without any such mutation, the probability that it will form is effectively 100%—the developmental system achieves this outcome very faithfully. But in mutant animals the probability may only be 30% or 50%, depending on the line. This is therefore a clear and stark illustration of the fact that intrinsic developmental variation can lead to quite distinct phenotypic outcomes, even from the identical starting genotype. The same effect is observed in humans, where phenotypes like agenesis of the corpus callosum are often not concordant between MZ twins—it may be absent in one twin and nearly normal in the other.

A number of psychological traits that show quite distinct outcomes may similarly be affected to a large degree by developmental randomness. One of these is handedness. Humans, like many mammals, show a strong preference for using one of their hands in complex manual tasks. What is unusual in humans is that the choice of hand is systematically biased, with around 90% of people being right-handed. Hand preference emerges or consolidates in infants around two years of age but does not seem to be instructed by experience—quite the opposite, in fact. While many naturally left-handed children have been forced to write with their right hands, their natural preference for using their left hand for other tasks remains strong. This suggests that handedness is a highly innate trait, even though it requires a period of maturation to fully emerge.

However, left-handedness is only partly genetic. Having one or both parents be left-handed increases a child's chances of also being left-handed, but only from a baseline of 10% to about 15% or 20%. Twin studies suggest that the overall heritability of the trait is only about 25% (indeed, MZ twins are still often discordant for left-handedness). There is no effect of the shared family environment (or upbringing) in such studies, meaning that the remaining variance in who becomes left-handed is attributed to the "nonshared environment" term—the term that can include effects of developmental variation.

Given how refractory hand preference is to outside influence, these findings strongly suggest that handedness is one example of a trait that is highly innate, despite being only partly genetic, largely reflecting instead the outcome of randomness in brain development. We will see in chapter 9 that sexual orientation may show this same pattern.

A ROLLING STONE

Many years ago, the famous developmental biologist Conrad Waddington came up with a visual metaphor that captures these kinds of "stochastic" processes in development (i.e., ones that follow a random probability distribution) and the way they can lead to quite variable outcomes (see figure 4.6). In this scheme, there is a little ball rolling down an undulating

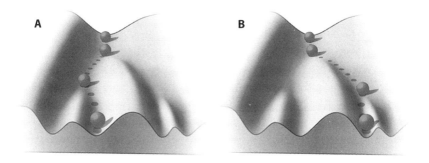

Figure 4.6 Randomness in development. The developing organism is represented by the ball, rolling down the "epigenetic landscape," the shape of which is determined by the individual's genetic makeup. At certain stages, very small differences in internal conditions (noise) can deflect the ball into one developmental channel or another, sometimes resulting in very different phenotypic end points, even in genetically identical individuals (**A** vs. **B**). (Reprinted from K. J. Mitchell, "The Genetics of Brain Wiring: From Molecule to Mind," *PLoS Biol.* 5, no. 4 (2007): e113; modified from original by C. H. Waddington.)

landscape, like the side of a mountain. The ball represents the organism on its journey through development, starting with fertilization at the top, and ending with the possible outcomes of development at the bottom. The landscape is characterized by various ruts and valleys that the ball may be channeled into and which lead to different outcomes.

As the ball rolls down it will come to points at the start of these valleys where its trajectory left or right will be affected by tiny random variations. If you were to let the ball go 100 times and see what happens, it might end up entering a hypothetical valley on the left 70 times and one on the right 30 times. But those probabilities will differ across individuals as the precise shape of each person's landscape reflects that person's own genome.

Imagine two possible outcomes that represent whether a person is right- or left-handed. For one person, the valley leading toward right-handedness as an outcome might be very deep and the entrance to it very wide, so that the ball is very likely to roll into it at the top. If we ran the ball down 100 times it might only end up in the left-handed valley 1 or 2 times. For another person, say someone who actually is left-handed, the landscape might be different such that this valley is easier to get into and the ball could end up there 10 or 20 times out of 100.

Instead of handedness we might think of clinical outcomes, like whether a person ends up with epilepsy, or autism, or schizophrenia. The inheritance of each of those conditions is also probabilistic—what is inherited is a genetic risk or predisposition, but whether or not the individual actually develops the condition is affected by nongenetic factors, with developmental variation likely playing a key role. For example, if one of a pair of MZ twins has schizophrenia, the chance that the other one will too is about 50%. Despite inheriting the same degree of risk, the outcome may thus be clinically much worse for one twin than the other. However, if you look at the offspring of such pairs of twins, their rates of schizophrenia are identical, whether or not their parent actually had the clinical condition. The risk is passed on, regardless of whether the person actually developed the condition.

You can use the same metaphor for more quantitative variation in traits like brain size or structural connectivity, if you imagine the ball rolling out onto a flatter plain at the bottom, representing a more continuous range of outcomes.

Waddington called this metaphor the "epigenetic landscape," based on the Aristotelian term "epigenesis," which refers to the processes of emergence or development of an individual. (This is not to be confused with the modern molecular biology term "epigenetics," which refers to a specific mechanism of regulating gene expression.) The epigenetic landscape nicely captures both the genetic variation between people and the opportunities for random variation to affect the outcome of each run of development. But it does much more than that—it also dramatically illustrates a key concept in the way that development works—that it is self-organizing.

In principle, there are an infinite number of ways that development could proceed—an infinite number of states of gene expression that any cell could be in and an infinite number of ways to arrange such cells in an organism. But the actual developmental pathways open to an organism are highly constrained by all the feedback interactions and control systems encoded in the genome. Only a limited number of global states or profiles of gene expression will be stable, because of all these interactions. Cells can become skin cells or muscle cells, but not something in between. And they can make a heart or a liver but not some weird or nondescript mass of cells. (What happens in cancer is that these cross-regulatory and feedback systems are disrupted or short-circuited in some way, releasing the normal constraints on cell and tissue fates.)

These developmental mechanisms have been shaped by millions of years of evolution with a single goal—making a viable organism (that will also have viable offspring). The self-organizing rules and protocols embedded in the genome thus ensure that the phenotype of the developing organism is channeled through a series of predictable stages toward that desired outcome. Waddington called this process "canalization" and conceived the ruts and valleys of the epigenetic landscape as a way to illustrate it.

ROBUSTNESS

In engineering terms, the system is robust. It is not enough to evolve mechanisms for self-organization that work only under optimal and highly defined conditions. In the first place, we have seen that molecular and cellular systems are unavoidably noisy—any system that relied on very precise molecular parameters would thus be hopelessly overspecified. In addition, developing organisms will often have to deal with fluctuations in the environment—for example, in maternal nutrition or changing maternal physiology during stress or infection—which can alter fetal biochemistry and cellular physiology.

The best that evolution can do is thus to make sure that there are enough redundancies and feedback systems encoded in the developmental program to deal with all these potential variables. But we must remember that there are limited resources available—there is after all a metabolic cost to building in all these fail-safe mechanisms, just as there is a financial cost in doing the same in engineering projects. Evolution rarely tolerates excess spending. These systems have therefore not evolved to be perfect—just good enough, under the kinds of conditions typically encountered. Under most circumstances they can accommodate and buffer the noise and any environmental fluctuations and generate an acceptable outcome. There may be some wiggle in the final result but this kind of quantitative variation is tolerated within the typical range.

There is, however, an unexpected consequence of the way that developmental systems are designed, which is a paradoxical fragility to certain kinds of perturbations, especially mutations in developmental genes. The robustness that evolved to buffer noise and environmental variables means the system can also absorb the effects of many mutations affecting components of the developmental program. But not all of them.

Some genes have such a crucial position in the regulatory architecture of developmental control systems that even minor mutations in them can have surprisingly large consequences. When this happens, development can end up being channeled down quite a different route—one of the alternate channels in Waddington's epigenetic landscape that leads to a different outcome. This represents a new trajectory and a new stable state of the system that is only uncovered when some of the normal protocols and regulatory relationships are compromised.

As we have seen with a few examples above, these effects are typically probabilistic. Mutations tend to not just change the outcome from one phenotype to another, but also increase variability. The reason is that the gene products that are affected do not just have their own specific roles to play—they also contribute generally to the robustness of the entire system. Messing with any of these components means the effects of noise are now less easily buffered, the possible states of the system are less constrained, and the outcomes become less predictable.

This is important not just for understanding the variability associated with genetic disorders—it is also important for all of us. There are no wild-type humans. We all carry thousands of minor genetic variants and typically 100–200 major mutations. So none of us has a developmental program that is as robust as it could be. If you or I were cloned 100 times, the result would be 100 new individuals, each one of a kind.

VARIATION IN VARIABILITY

This brings us to the final wrinkle in this story. The degree of robustness in individuals depends on their mutational load, which varies across people in many ways. First, some of us simply carry more major mutations than others—the average is somewhere around 150, depending on how they are defined, but there is substantial variation around that. Second, the particular set of mutations will affect more developmental genes in some people compared with others. And finally, the specific *combination* of mutations and of more common genetic variants that each of us carries may cause a greater or lesser impact on the developmental program.

This means that developmental robustness—or its converse, developmental variability—may itself be a genetic trait that varies between people, correlated with the effective mutational load. I mentioned earlier that you can see the effects of developmental variability by looking at facial asymmetry, but some people are more asymmetric than others. If you take many measures of different body parts and facial features you can get a composite measure of asymmetry in any individual. Twin studies have shown this trait is at least partly genetic, with a heritability of about 30%. Indeed, if you look at pairs of MZ twins, there is a correlation between how similar they are to each other and how symmetric each twin is.

So, if we think about our little cloning experiment again, we can expect different results for different people—your 100 clones might be more or less variable than mine. Your genome might encode a broader or narrower range of possible outcomes than mine, depending on our relative mutational load and degree of developmental robustness.

The excellent science fiction movie *Gattaca* envisions a dystopian future where successive generations of genetic screening have removed most deleterious mutations from the population.[1] This generates extremely genetically "fit" people, embodied by characters played by the ridiculously good-looking Jude Law and Uma Thurman. The protagonist of the film, portrayed by Ethan Hawke, was bred the old-fashioned way, however, and carries a greater genetic burden. He's no slouch himself in the looks department, to be fair to him, but the filmmakers did a good job in casting because he is clearly not as symmetric as Law or Thurman. Indeed, ratings of physical attractiveness are consistently correlated with facial symmetry (not completely of course, but partly). This makes evolutionary sense if symmetry is a reliable indicator of developmental robustness and, thus, genetic fitness—we should be attuned to displays of genetic fitness as markers of attractive mates.

Variation in developmental robustness may also show up in one of the most important traits in humans—intelligence. There is a consistent positive relationship between higher facial symmetry and higher intelligence. Again, this is only a partial correlation, with a correlation coefficient on the order of 0.12 to 0.20. But it's enough to suggest that higher intelligence may reflect in part a more robust developmental program, one that is better able to direct an outcome closer to optimal in terms of neural organization and efficiency. We will look at this in much more detail in chapter 8.

As well as having effects by itself, the degree of developmental robustness of a person's genome may also modulate effects of other factors, including how a developing embryo or fetus reacts to external challenges, like maternal stressors, as well as how it copes with the effects of specific mutations. Some individuals may be more resilient to such insults because their developmental program is robust enough to buffer their effects. This kind of very general effect of mutational load on robustness may partly explain why some serious single mutations can cause quite different clinical effects across individuals with different genetic backgrounds. More on that in chapter 10.

[1] *Gattaca*, directed by Andrew Nicoll (Los Angeles: Jersey Films, October 24, 1997).

THE END OF THE BEGINNING

Throughout this chapter, I have been discussing the unique "outcome" of development in any individual, as if this were a fixed and static endpoint. Of course it is not. The molecular and cellular processes described above set up the initial patterns of brain organization and connectivity, but these are merely the first steps in brain development. The brain comes prewired, but not hardwired. In the next chapter we will look at how processes of brain plasticity lead to refinement of neural circuitry in response to experience, and how such effects, far from evening out initial differences between people, can instead amplify them.

CHAPTER 5

THE NATURE OF NURTURE

The debate about the relative contributions of nature and nurture to our psychological makeup is classically framed as a battle between these two forces, rather than, say, a collaboration. In recent times, this has turned into a proxy war, with genetics on one side and brain plasticity on the other, lately allied with the shadowy forces of "epigenetics." If the brain can change itself, and if we can turn our genes on or off by our own behavior (which is what some proponents of epigenetics rather nebulously claim), then it seems we could reverse the arrows of causation—our psychology could dictate our biology, rather than the other way around.

Under this scheme, nurture—whether this refers to parenting, experiences, or our own conscious psychological practices—can trump nature. It can overwrite the innate differences in our brains that arise due to genetic and developmental variation. In fact, what tends to happen is just the opposite—initial differences tend to be amplified due to the self-organizing processes of brain development and the fact that individuals select and construct their own environments and experiences largely based on innate predispositions. This is a radically different conception, where the processes of brain plasticity—the supposed instruments of nurture—align with nature instead.

BRAIN PLASTICITY

Our brains come prewired, but they are not hardwired. At birth, we have extensive individual differences in brain wiring due to genetic and developmental variation. But brain circuitry is, at a microscopic level at least, highly plastic. In fact, you could say that the brain's main job is to change itself—that is how it reacts to the environment and how it stores memories of experiences, tracking the statistical patterns in the world, mapping out causes and effects, tagging outcomes as good or bad for future reference. Anything that we have learned has a physical substrate somewhere in the brain—a change in synaptic connections between some neurons, which will alter our response to the relevant stimulus or situation when we encounter it again.

Learning from experience obviously affects our behavior—that is why we do it. But does it affect our behavioral *traits*? We can learn to recognize certain situations, to predict outcomes of various possible actions, and weight them according to short- or long-term goals. But underlying those decisions are certain predispositions, which explain why different people weight various options differently. One person may value possible rewards more positively or possible punishments more negatively than another, a person may be more or less risk averse, more sensitive to possible threats, better able to inhibit immediate impulses, better able to defer short-term goals in the service of long-term ones, and so on. The plasticity we see in adult brains lets us learn information about the world, but there is little evidence it can change those underlying predispositions.

But what about in children? Could similar processes active during the extremely extended period of brain maturation in humans help shape such predispositions, by changing wiring in more profound ways, dialing up or down the responsiveness of various control circuits as an adaptive response to types or *patterns* of experiences?

THE IMPACT OF EARLY EXPERIENCE

There is a large body of evidence supporting the idea that extreme experiences in childhood—of neglect, abuse, or maltreatment—can indeed have long-lasting effects on people's psyches. For example, studies of children raised initially in orphanages, particularly ones where they suffered serious neglect, have found correlations with a variety of psychological traits—such children show reduced "developmental quotient" (basically, child IQ), indicating a delay in cognitive development, deficits in attachment, more behavioral problems, attentional difficulties,

and problems with peer interactions. Some of these effects can be ameliorated or reversed when the children are later adopted into a family
environment, though others may remain. Typically, those adopted earlier show better outcomes than those who stay longer in an institution.

These are average effects, of course, and there is evidence of substantial heterogeneity in how individual children react, with some showing more resilience to the negative effects of institutionalization than
others. In addition, most of the relevant studies have examined these
people while they are still children, so the persistence of such deficits
into adulthood remains unclear. A further difficulty with such studies
is in identifying large enough numbers of children with these circumstances to draw firm conclusions.

An alternative approach has therefore been to assess the effects of
maltreatment across samples of the wider population. Many such studies have fairly consistently found that exposure to "early life stressors"—
including various indicators of emotional trauma, neglect, or physical
or sexual abuse—correlates with defects in emotional processing and
regulation (with heightened negativity and decreased positivity); in-
secure, disorganized attachment behavior; difficulties with peer interactions (characterized by high withdrawal or aggression); and a range of
effects on personality traits (such as lower agreeableness, conscientiousness, and openness to experience, as well as higher neuroticism) at later
ages. These differences in psychological traits are reflected in increased
rates of mood and anxiety disorders, attention deficit/hyperactivity disorder (ADHD) or conduct disorders, post-traumatic stress disorder,
substance abuse, suicide attempts, and other psychiatric presentations.

Again, these are average effects, and not all children react in the same
way to such experiences. But on the face of it, these studies do suggest
that early life experiences *cause* differences in later behavior. They don't
actually show that, however—all they show is a *correlation* between
early experiences and later behavior. There are actually a number of
other possible explanations for such a correlation. First, a child's own
early patterns of behavior might make that child more likely to experience stressful life events. This is going to sound like blaming the victim,
which is not what I intend, but it is certainly conceivable that a child
who is, say, naturally aggressive or has behavioral "problems" (meaning
the child's behavior is a problem for other people), might attract more

negative treatment from parents. In fact, there are good data to suggest this kind of effect may well be a contributing factor to maltreatment (without in any way excusing the maltreatment), and would obviously also correlate with the child's later behavior.

Another possible explanation that is a bit more subtle is that the parental behavior and the child's behavior could both be independent manifestations of shared genetic effects. You could certainly imagine how many of the traits that are the supposed outcomes of maltreatment, which are all at least moderately heritable, might themselves predispose parents to maltreat their children. Without controlling for genetic relatedness, such studies cannot rule out this possible confound.

On the other hand, controlled experiments in rats and mice have shown that early life stressors of one kind or another can lead to behavioral differences later on. There are even data showing that the activity of specific genes involved in the stress response can be altered in a long-lasting way by early stress experiences (through what are known as epigenetic changes, which affect gene expression). There is thus supporting evidence from animals, and possibly even some mechanistic insights, into how early experiences can have long-lasting behavioral effects.

So, for now, let us set aside the concerns about possible genetic confounds and take at face value the studies claiming that extreme differences in early childhood experiences can have long-lasting psychological consequences. One might extrapolate from those observations that less extreme differences in nurture could also contribute to psychological traits, just on a more subtle scale, explaining some of the variation across the normal range. However, the twin, adoption, and family studies we have been considering explicitly argue strongly against such an effect. They consistently show no or very little contribution to variance in psychological traits from variation in family environment.

WHY DON'T WE SEE EFFECTS OF UPBRINGING?

If differences in nurture *can* affect our psychology, why don't they? Or, to be more accurate, why don't we see the signatures of such effects in twin studies? There are several possible explanations. First, it could be that only extreme experiences have any effect—it may require serious neglect or abuse to derail the otherwise robustly innate trajectories of development. If such experiences are rare across the population—or if they are undersampled in the populations studied (a real possibility in adoption studies especially)—then they will not contribute much to the overall variance in a trait. As a result, the "shared environment" term in analyses of variance in twin studies will remain small. Remember, these analyses don't show us everything that *could* contribute to variance, they just show us what actually does, in the particular population under study.

Second, our early experiences might indeed be important sculptors of our psyches, but this effect may be driven more by experiences outside the home. Variation in such experiences would contribute instead to the "nonshared environment" term in twin and family studies, which we know is sizable. Sadly, we know of many such potential sources of abuse or maltreatment outside the home, which are indeed likely to leave lasting psychological scars. And it is possible that such effects extend across the more typical and less traumatic range of experiences, contributing to variance in psychological traits across the entire range. But assigning primacy to nonfamilial interactions seems like a case of special pleading. To accept this as a valid explanation, we would have to believe that experiences with peers, teachers, coaches, or other non–family members can have long-lasting effects, while experiences with our parents and family members cannot. And we would further have to conclude that being brought up in the same family has no systematic correlation with the types of experiences individuals have outside the home. These assumptions violate both common sense and common experience and there is little or no evidence to support them.

A third possibility is that our experiences do affect our traits, but these effects are highly dependent on our genetic makeup. People may differ in how vulnerable or resilient they are to trauma or maltreatment, with the result that such experiences will have lasting psychological effects in some people, but none at all in others. This seems a highly likely scenario—in fact, we know that people do differ in resilience and we even know that this is a heritable trait. But can this explain why we don't see a shared family environment effect in twin studies? Not really, because such studies look at the average effect of family environment across all individuals in the sample—even if the magnitude of this effect varies, unless it averages out to zero (which would mean it would actually have to have effects in one direction in some people and effects in the opposite direction in others), we should still see some signature of it if our samples are large enough.

This brings us to the final possibility for why we don't detect an effect of nurture or experience in twin and family studies—that we are looking in the wrong place. Such studies typically think of environmental or experiential factors as things that happen *to* an individual. But, actually, many experiences and environments are actively chosen *by* an individual. Variation in such experiences may indeed shape our psychological traits, as a *mechanism* of change, but the *source* of such variance may ultimately be the innate differences between people. Their effects will therefore show up on the "nature" side of the ledger. An aptitude for music, for example, may lead people to pursue musical training, which will only serve to increase their musicality.

Rather than overriding or flattening out their effects, processes of brain plasticity may reinforce and even exaggerate the widespread initial differences that arise due to both genetic and developmental variation. As we will see below, this kind of effect happens as people select and construct their own environments and experiences throughout their early life. But it actually starts even before that, prior to anything we could call "experience"—prior to birth, in fact.

THE SELF-ORGANIZING BRAIN

Brain development does not end once all the cells are in the right places and all the synaptic connections have been made. The developmental program encoded in the genome generates a detailed, but still coarse, sketch of the brain's wiring diagram. This is followed by extensive refinement of these circuits, with synapses added or removed depending partly on an ongoing genetic program of maturation but also partly on the patterns of electrical activity that flow through them. These patterns are driven by experience after birth, but before birth they are driven by spontaneous activity that arises in neurons as soon as they make connections with each other.

The brain uses this "beta-testing" phase to fine-tune and optimize its circuits, to coordinate levels and patterns of activity across interconnected structures, and to streamline what is initially an exuberant and expensive set of connections. Synapses are modified—strengthened, weakened, added, or pruned—based on two simple principles: "cells that fire together wire together" and "use it or lose it."

This happens through multiple mechanisms. One of these involves a change in the biochemical makeup of the synapse in response to previous activity. The sensitivity of a downstream neuron to the release of neurotransmitter at any given synapse depends on how many molecules of receptor proteins it has on its surface, as well as the levels and biochemical states of myriad other proteins in the dendrite. If a synapse has been repeatedly activated, the levels of neurotransmitter receptor molecules may be increased, thus making the neuron more sensitive to subsequent activation. This may be followed by growth of new synaptic connections between the neurons in question, forming an even longer-lasting increase in connection strength (see figure 5.1).

These kinds of changes—which can go in the opposite direction too, to weaken or prune synapses—are thought to underlie learning and memory. When played out across whole networks of neurons, they provide a record of previous experience—for example, that a certain stimulus is associated with a certain outcome. During development, the same mechanisms are used to fine-tune brain circuits.

These principles have been worked out in the study of major sensory systems, especially the visual system. At a gross level, inputs from the retina are required for the rest of the visual system in the brain to develop normally. Retinal ganglion cells form the output layer of the retina, sending their axons along the optic nerve to a number of regions of the brain, including one called the thalamus, or inner chamber. This is a small, centrally located hub region that acts as a relay for sensory information from the peripheral sense organs. It is divided into many discrete regions, each devoted to processing different kinds of information. The visual center of the thalamus receives inputs from the retina, and, in turn, projects to what will become the primary visual cortex—the first stop in a hierarchy of cortical areas devoted to vision.

If the eyes are absent or if the patterns of electrical activity in the retina are altered, then the circuitry of the visual thalamus and visual cortex will be affected. Indeed, in congenitally blind individuals, what would normally become visual cortex develops with altered connectivity and

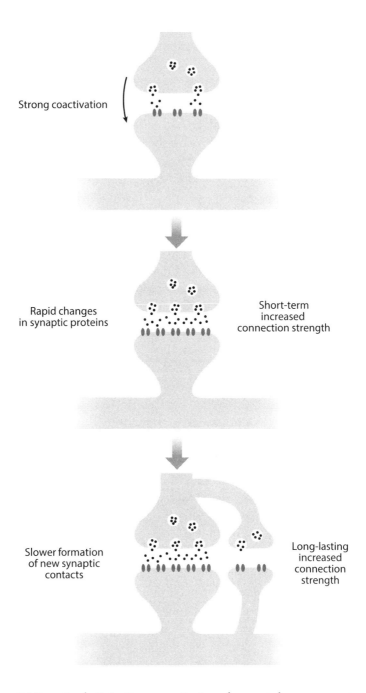

Figure 5.1 Synaptic plasticity. Strong coactivation of connected neurons may increase the strength of the synapse, so that the output response to the same signal becomes greater. This is accomplished by short-term changes, such as an increase in neurotransmitter receptor proteins, or by longer-term growth of additional synaptic connections.

becomes responsive instead to auditory or somatosensory (tactile) information. The reactive expansion of cortical territory devoted to these other senses likely underlies the increased sensitivity to sound or touch exhibited by many blind people.

Similar effects are also seen with less drastic manipulations. We found in my own lab that a reduction in the number of axons projecting from the visual thalamus to the visual cortex (due to a mutation that alters the guidance of these axons so that some of them get lost along the way) leads to a concomitant reduction in the area of primary visual cortex. This kind of sensitivity to levels of inputs seems like a highly general phenomenon, extending to higher-order areas as well, which receive inputs from lower-order areas. Initial differences in the development of one area or the wiring between areas thus have cascading effects on the subsequent development of brain-wide networks.

At a more subtle level, the patterns of connectivity within a particular area of the cortex—the specific microcircuits that develop to process different kinds of information—are also dependent on the appropriate inputs and patterns of activity. Even prior to birth or prior to eye opening, waves of electrical activity wash across the retina, such that neighboring neurons will tend to be active around the same time. This is crucial for refining the map of the visual world that is conveyed by retinal projections to visual centers in the brain. This mapping is initially established through the actions of proteins expressed in matching gradients across the retina and its targets. These molecules direct the termination of retinal axons at appropriate positions on the target region, such that neighboring retinal ganglion cells terminate on neighboring target region cells, and the overall topography of the retina is smoothly mapped out across the surface of the target region. The initial map is somewhat coarse, however, and the waves of activity across the retina are required to refine it, making sure that those connections that faithfully maintain the visual image on the retina are strengthened and any stray axons that would confuse visual processing are pruned away.

After birth, the processes of brain plasticity continue to sculpt and refine the circuitry of sensory systems, but they are now driven not by correlated patterns of spontaneous activity, but rather by the statistical regularities of the world we experience. This enables the system to develop expertise for the kinds of things it normally encounters, and that are important for our survival, though in the process it loses flexibility to process things outside the scope of its typical experience.

Early experience, during what are called "critical periods," is crucial in developing this perceptual expertise. If sensory experience is absent or degraded in some way—for example, by congenital cataracts or by conditions that impair hearing in early infants—then, even if those conditions are later corrected, the full range of normal visual or auditory perception may never develop. The system must have normal experience during the critical period (the first few years, in these cases) in order to optimize its circuitry.

This plasticity thus allows the system to adapt itself to the regularities of the organism's experience. All that information doesn't have to be built into the genome—into the developmental program that directs the initial wiring patterns—because it exists in the world. Instead of possibly overadapting to a specific environment, evolution has programmed in flexibility. The final, fine level of adaptation is achieved by the organism itself, on an individual basis, not on a species basis.

A SELF-TERMINATING PROCESS

The process of refinement by experience is, by its very nature, self-terminating. Connections that are reliably driven by sensory stimuli will be strengthened and those that conflict with patterns of experience will be weakened or even pruned away. That process thus changes the patterns of activity that arise in response to the next stimulus, biasing them toward one pattern and away from another, in turn *further reinforcing* that bias by the same processes of plasticity. Eventually, through this positive feedback, you will get a system that is very good at processing certain types of stimuli—the ones we encounter reliably and that matter to us—but that has lost its capacity to learn to discriminate other types of stimuli. Indeed, the biochemical processes of plasticity that underlie wholesale activity-dependent refinement are actively turned off in the brain after a certain stage (at different times for different systems), consolidating the now-optimized circuitry and closing off potential for further change.

A good example of this is language perception. As infants are exposed to a primary language, they develop expertise at categorically recognizing the characteristic phonemes, or speech sounds, of that language. For example, native English speakers become adept at distinguishing between the sounds of "b" and "v," or "r" and "l." Spanish speakers, by contrast, may not distinguish so readily between "b" and "v" sounds, while Japanese speakers may have difficulties hearing a distinction between "r" and "l." Amazingly, EEG (electroencephalogram) recordings show that the auditory regions of the brains of Japanese infants make that distinction just as well as infants exposed to English as a first language. But that ability is *lost* over time. The process of developing expertise to sounds in one language eventually closes off the ability to distinguish between sounds that are not heard as often or between which making a distinction has never been important. The phonemes "r" and "l" thus literally sound the same to Japanese speakers, in the way that the tonal subtleties of Cantonese may be completely lost on native English speakers. It is this loss of flexibility that explains why we lose, after a certain age, the ability to learn a second language without a telltale foreign accent.

The development of sensory systems nicely illustrates the influence of activity-dependent plasticity on the refinement of neural circuitry. However, the role of activity in these systems is normative; that is, there are not typically differences in the *quality* of experience that contribute to variation in the outcome. Apart from experiments where kittens are raised in a visual environment consisting only of vertical lines, or goldfish are raised under strobe lighting, the natural environment itself is not typically a *source* of variation. However, there certainly can be variation in the subjective experience of that environment, as in children born with cataracts, for example. In such cases, the source of that variation is *the individual*, not something in the world. The mechanisms of plasticity then serve to reinforce and even amplify those innate, intrinsic differences, exaggerating the range of phenotypic differences without reflecting variation in the environment per se.

HABITS OF MIND, HABITS OF BRAIN

The processes that refine connectivity in sensory systems apply just as much in other parts of the brain, including in circuitry that mediates behavior. As an organism encounters the world, it has a wide range of options for how to respond. These depend on how it appraises the situation: Are there any threats? Is there anything to eat here? Can I mate with that? Just as in the case of sensory perception, it is the *subjective experience* of the environment that matters. A certain situation may contain an objective threat, but one individual may *feel it* as more threatening than another does.

It is those so-called "affective" or emotional states that drive initial behavioral responses in young animals or human infants. Prior to any knowledge or experience of the world, innate instincts program behavior by assigning a *value* to different states. Pain is not just a signal about damage to some body part—it is *painful*! It commands attention and demands a response. Hunger is not just a signal that you need food—it feels *bad*. When it reaches a certain level, it can't be ignored—it entails a strong urge to seek food. And that food is not just nourishing—it is rewarding. It feels *good*. Affective signals thus tag information about physiological states with value and thereby drive behavior, leading the organism to try to maximize the good states and minimize the bad ones.

With experience, the same thing happens with information about the outside world—various elements of it are tagged as things that could be good or bad for the organism. And various possible actions are also tagged with value based on prior experience and predicted outcome—the nervous system's way of asking "How'd that work out for you last time?" In all these things, it is the affective signals that guide learning—these are the signals that tell the system not that the outcome of an action was X, but that X was good or bad.

In the process, these signals also strongly influence which aspects of experience an individual learns from. We don't need to learn from everything we encounter in the world or even from every action we take—we only need to learn about things that matter. Most things in the world are neutral, as far as we are concerned—we needn't really care about them and it's not adaptive to modify our brains or our behavior based on them. We only need to learn from things that were good or bad for us, which we know only because they *felt* good or bad, *to us*—because they were tagged with subjective, affective value.

As we explore our environment, we start to build up a record of contingencies in the world—of things that are commonly associated, of causes and effects, decisions and consequences. The imprint of these experiences is laid down in circuits of the cerebral cortex (and interconnected structures such as the thalamus and basal ganglia), based on exactly the same kinds of plasticity rules that apply in developing sensory systems—strengthening connections that are used a lot and pruning away those that are underused. Through these processes, recurrent *patterns of activity* will get reinforced—the connections between all the neurons that make up an active pattern get strengthened, because they are all active at the same time or in sequence. This means it becomes easier to reactivate that pattern on subsequent occasions. The more times an organism's brain enters that state, the more likely that state becomes in the future—it starts to form a habitual response, at the neural level. Conversely, connections that are rarely used together will get pruned away, thus reducing the possible number of states that the system can enter in to, further biasing the system toward those states that have been strengthened.

But that doesn't happen all the time—the processes of plasticity are gated by the affective processes that tag our experiences with value. These systems control overall levels of arousal (something important is happening!) and direct attention toward specific elements in the environment (that thing could kill me!) or toward outcomes of actions that matter (that didn't turn out well—I should definitely not do that again!). These signals are conveyed to the cortex by the release of neuromodulators from circuits from the brainstem, hypothalamus, and other "primitive" areas that mediate affective states. These neuromodulators act on the synapses in cortical networks, controlling the processes of synaptic plasticity—these only kick in under circumstances when arousal is high and attention is focused on the relevant elements of the situation.

In this way, we learn from our experiences and adapt to our environment. *But we don't all experience the world in the same way.* Innate differences in the subjective weights of these affective states in different individuals will strongly bias activity-dependent learning in higher-order behavioral circuits. What different people find salient differs from the outset—some people may weight threats more heavily, they may be more risk averse, they may find positive outcomes more subjectively rewarding, or negative outcomes more subjectively aversive. Because the weight of these signals is the thing that gates learning, initial differences will tend to get reinforced and exaggerated through these processes (see figure 5.2).

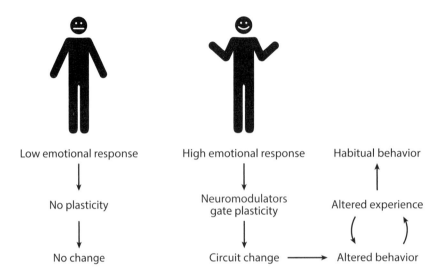

Figure 5.2 Subjective experience drives plasticity. Innate differences between people in emotional responsiveness to particular types of stimuli regulate synaptic plasticity and can thus be amplified by experience and lead to differences in habitual behaviors.

If you find rewards more rewarding—if they feel more positive to you—then this will not only bias your initial behaviors, it will also mean that you will reinforce brain states that led to rewards more than someone else would. The next time you encounter a similar situation, you will weight those possible actions more heavily than someone for whom the same rewards are just not valued as highly. This creates a positive feedback loop, where those patterns of activity become habitual—for you. But for other people, the same sort of experiences might lead them to adopt quite a different habitual response, based on differences in their initial responses and what aspects of their experience they tend to learn from.

Pretty soon, you may be not just reacting to situations you find yourself in but proactively making decisions that will put you in one type of situation versus another—choosing ones you find rewarding and avoiding ones you find unpleasant or aversive. This leads to a higher level of reinforcement of innate tendencies—not just a passive amplification of typical responses, but also an active selection of situations, environments, and experiences to which an individual has self-adapted.

CHOOSING OUR ENVIRONMENTS

There are multiple ways in which people's experiences are either passively affected by their innate genetic makeup or actively selected based on their psychological predispositions (see figure 5.3). The first is really an indirect effect. It comes from the fact that children are typically raised by their biological parents, with whom they share many genetic variants. Those variants may affect both the child's behavior and the parents' behavior, in ways that may interact. For example, a naturally anxious child may also have naturally anxious or overprotective parents, whose behavior will then tend to validate and exaggerate the child's own tendencies. Naturally aggressive children may be more likely to have parents who are also aggressive, which may thus amplify the child's predispositions through heightened conflict.

This kind of interaction is thus another potentially major confound in sociological studies that do not take shared genes into account. Not only can correlated behaviors between parents and children be caused independently by shared genes (such that the parenting behavior does not cause the child's behavior), they can also show a more complicated interaction, where the shared genetic propensity in turn amplifies the

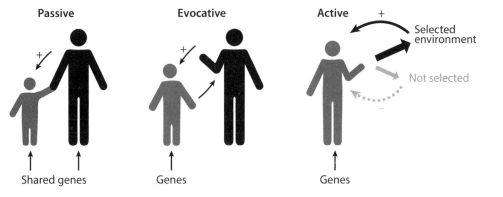

Figure 5.3 Genetic effects on our experiences. *Passive*: a child's experience may be affected by shared genetic variants that influence both its own and its parents' behavior. *Evocative*: a child's innate traits may evoke specific types of responses in parents, teachers, or peers. *Active*: children tend to seek out experiences that suit their own psychological makeup, thus reinforcing those traits.

behavior. In the latter case, the *mechanism* of change includes a real differential effect of parenting, but the ultimate *source* of variation in behavior is still genetic.

Another important way in which a child's innate tendencies will influence the nature of the experiences that it is exposed to is by evoking different kinds of responses from the people around it. A difficult or willful child may evoke more negative, disciplinary reactions from parents or teachers, for example, even to the point of receiving greater corporal punishment than a more placid child who is more eager to please. An agreeable, conscientious child will, by contrast, receive more encouragement and positive reinforcement, thus rewarding these behaviors in a classic positive feedback loop. This is not just conjecture—twin and adoption studies have demonstrated that much of the variance in how parents treat their children is indeed driven by the child's own behavior and genotype.

These kinds of passive effects due to shared genetics and evocative effects based on people's reactions play important roles, especially in shaping early experiences. As children grow and develop more autonomy, they also begin to more actively select their own experiences and construct their own environments. A cautious child, who innately weights risks very heavily, may consequently avoid risky situations and thus not develop the expertise of managing risks and the confidence that it engenders. That child's reckless sibling, by contrast, who may discount risks and perhaps find novelty more rewarding, will have a different, probably more varied, set of experiences, which will likely reinforce these innate tendencies.

Again, longitudinal twin studies provide direct evidence for these kinds of effects. Variation in many aspects of people's life experiences—from the type of parenting and family environment they experienced, to their interactions with peers and social support, even marital quality and number of stressful life events experienced—all show substantial heritability. That is, pairs of MZ twins have significantly more similar life experiences than pairs of DZ twins.

In many cases, all these kinds of effects may act together. For example, a naturally intelligent child may also have parents who are more academically inclined and who therefore encourage application to schoolwork (a passive effect). The child's performance at school will likely be more rewarding due to more positive feedback from teachers (an evocative effect). And the child itself may choose to spend more time on academic activities due to innately finding them interesting and rewarding (an active effect). An initial difference in intelligence between two children could thus be amplified through all these factors and reinforced through what becomes an unequal experience of education.

Evidence for the importance of this kind of mechanism comes from a striking and consistent finding from twin studies of intelligence: the heritability of this trait *increases over time*. When it is assessed in young children, about 50% of the variance is associated with genetic differences, while the shared family environment also makes a sizable contribution— 30%–40%. However, when assessed in adults, the effect of shared family environment goes to zero, while the heritability increases to 80% or more. This suggests that early family environment does indeed make a difference to the pace of early cognitive development, which childhood IQ tests may measure—however, this does not seem to have a lasting effect on final cognitive ability. By contrast, the early genetic effect is amplified over time, such that MZ twins show higher concordance in absolute IQ measures in adulthood than they do as children.

This sort of effect is probably even more obvious for cases of musical talent or innate athletic ability. Children with such talents will often quite naturally be strongly encouraged to pursue the kind of training that will develop these skills, while those with a tin ear or two left feet may be directed to other pursuits. And those with a natural talent for an activity will likely find such pursuits more rewarding, thus contributing to the positive feedback loop, which will exaggerate initial differences between individuals.

It is important to remember that those initial differences—what we've been calling "nature"—come from both genetic and developmental variation. Genetics sets the starting point, but the outcome of development from any individual run will still be quite unique. A clever study in mice illustrates this point and also shows that it is not just the absolute levels of a trait that matter (high or low); it is sometimes the relative levels across interacting individuals (higher or lower) that influence outcomes.

This study followed a large cohort of mice from birth through to adulthood, as they lived and interacted with each other in a large, complex enclosure. All the mice were from the same inbred strain and were therefore genetically identical to each other. Like many social creatures, mice typically have quite strong dominance relationships. By following the outcomes of encounters between specific individuals, the relative rankings of each mouse could be determined. The striking observation was that at the start, when the mice were young, there were only minor differences in dominance apparent between them. But over time, it was the ones who started out with a slight advantage that eventually became the most dominant animals. That tiny initial difference was amplified by repeated encounters, because winning a confrontation increases relative dominance ranking and losing decreases it.

AMPLIFICATION BY CULTURE

We can also see this kind of amplification of initial differences writ large at the cultural level. This is particularly obvious in relation to sex differences. We will see in chapter 9 that there are many group average differences in cognitive and behavioral traits between males and females. (This means that though the distributions of values for a given trait vary widely for both males and females and overlap substantially between them, there is a significant difference in the *average value* between males and females—as with height, for example.) These include group average differences in interests and values, with males tending to be more interested in things or systems and females tending to be more interested in people.

These average differences are apparent at a cultural level, in terms of expectations about what men and women should be interested in generally, what kinds of subjects they should study at school, and what kinds of professions they are suited to. Indeed, some people would say that these cultural expectations drive the observed differences in the first place. Others argue that these differences are completely biological and the culture merely reflects them. In fact, both positions may be right. The cultural expectations may arise and persist due to real biological differences between males and females, but they may also contribute to a self-fulfilling prophecy, in that those expectations shape the experiences of boys and girls, offering them different types of experiences and opportunities, which will thus tend to amplify the initial biological differences.

CLOSING OFF OF POTENTIAL

Humans have an extremely protracted period of brain development. This is especially true for circuits that mediate behavioral control, such as the prefrontal cortex, which do not fully mature until a person's early twenties. Until that time, synapses are still being modified on a massive scale. This provides ample opportunity for these circuits to be shaped by experience and is likely one of the key factors in our ability to successfully populate what has been called the "cognitive niche." Rather than being adapted for specific environments, with a limited set of hardwired, instinctive behaviors, we have evolved cognitive flexibility and responsiveness, allowing us to adapt ourselves to our individual environments. Recurrent patterns are reinforced and habitual modes of behavior emerge. We gradually become ourselves.

But at some point we have to stop constantly becoming and just get on with things—important things like building a career or finding a mate. That means we have to consolidate the adaptations we have made and restrict further changes. We can't have runaway positive feedback loops forever—we have to maintain these neural configurations to remain ourselves. The periods of wholesale plasticity last considerably longer in behavioral and cognitive circuits than in sensory ones, but they still close as we reach adulthood. The plasticity processes themselves will have progressively narrowed the "degrees of freedom" of the developing brain, magnifying initial biases by both positive reinforcement and progressive elimination of connections mediating less-favored states. But the biochemistry of the brain also changes with maturation, so that mechanisms of plasticity and flexibility get replaced by mechanisms of stability and maintenance.

Far from flattening out or overwriting initial differences, our experiences thus tend to consolidate them instead. These processes drive what developmental neuroscientist Marc Lewis has referred to as "the growing inertia that each developmental path accumulates over time—the eerie manner in which developing humans become increasingly crystallized versions of themselves."[1]

---

[1] M. D. Lewis, "Self-Organizing Individual Differences in Brain Development," *Dev. Rev.* 25 (2005): 262.

CHAPTER 6

I, HUMAN

There are anywhere from 4,000 to 8,000 words in the English language that refer to personality traits—fairly stable aspects of people's characters that are useful shorthand descriptors and predictors of their behavior. That these words exist at all is testament to the fact that people really do have such traits—they really do behave in more or less characteristic ways that differ between individuals. However, there aren't really 8,000 of them—many of these words refer to effectively the same thing. For example, I might refer to myself as determined, resolute, unwavering, or single-minded, while others might refer to me as stubborn, inflexible, obstinate, or even pig-headed. While these all have slightly different nuances, they clearly all map to a similar underlying construct.

By analyzing the lexicon of personality words, many researchers over many decades have tried to abstract a core set of factors that encompasses most of the diversity. The idea is that there may exist a limited number of underlying psychological parameters that are mutually independent of each other and that manifest themselves through related personality traits.

Exactly how many factors there are and what they are remains a subject of debate, however. In the 1940s psychologist Raymond Cattell deduced a structure with 16 primary trait factors. Though he avoided naming them himself, they roughly correspond to the colloquial meanings of warmth, reasoning, emotional stability, dominance, liveliness, rule consciousness, social boldness, sensitivity, vigilance, abstractedness, privateness, apprehension, openness to change, self-reliance, perfectionism, and tension.

Several decades later, Hans Eysenck deduced only two major factors, which he called Extraversion and Neuroticism, though each of these had multiple facets contributing to it. For example, people high in Extraversion were sociable, risk taking, dominant, sensation seeking, active, and expressive. People high in Neuroticism were anxious, depressed, guilty, and had low self-esteem. Since these traits are independent of each other, you could have people high in both, low in both, high in one and low in the other, or more balanced. Eysenck later added a third dimension, which he called Psychoticism—people high in this trait were aggressive, assertive, manipulative, and egocentric.

There are multiple other schemes but the one that is most popular in personality research these days is a five-factor model, known as the Big Five. These include Extraversion and Neuroticism, pretty much as Eysenck defined them, as well as Conscientiousness, Agreeableness, and Openness to Experience. People high in Conscientiousness tend to be organized, efficient, dependable, and dutiful, but not very flexible or spontaneous. Those high in Agreeableness tend to be friendly, compassionate, cooperative, and helpful. And those high in Openness tend to be imaginative, aesthetic, curious, and inventive, but also unpredictable and unfocused.

Those are obviously just qualitative descriptors, but it is possible to assign a quantitative value to these traits based on responses on questionnaires. You can give people a list of statements and ask them how much they agree or disagree with them, on a five-point scale. These could include statements such as: I enjoy parties, I like to travel, I often feel anxious, I am very competitive, I like learning new things, I tend to obey the rules, etc. What is found is that the scores on some of these items tend to be correlated with each other. For example, people who enjoy parties tend to also like to travel and to be more competitive than average.

This suggests that some *common underlying factor* is contributing to the scores on those three items. If you combine the scores on those and related items, that gives you a score for that latent factor, in this case defined as Extraversion. This does not explain all the variation in those specific items, but it does account for the correlation between them. You can do the same for items that load on Neuroticism ("I often feel anxious") and the other factors of the Big Five, which all vary at least partly independently of each other. The actual numbers are completely arbitrary, of course, but this method does give you a way to rank people in a pseudoquantitative fashion.

There tends to be reasonable agreement in scores for individuals as rated by themselves or by other people who know them. And the scores tend to be fairly consistent if the same person is tested over separate sessions—test-retest reliability is around 0.7 (where 1 would be perfect agreement). That's not bad, but it does give an indication of how fuzzy these numbers are. The values also tend to be consistent for individuals over periods of many years. At least, there is pretty good consistency in *relative rank* between people over time, even though there are also typical changes that occur across ages. For example, teenagers tend to be higher in Extraversion (especially sensation seeking) and Neuroticism, while older people show higher levels of Conscientiousness.

These Big Five factors are also consistently observed across countries and cultures. Even though different nationalities are perceived as having different character traits (e.g., Germans as organized, Italians as hot-headed, etc.), these do not manifest at the level of basic personality dispositions. Mean values of the Big Five traits do not differ significantly between countries. If these different national character traits have any validity at all, they thus likely reflect cultural influences that apply on top of individual variation in more fundamental behavioral dispositions, rather than differences in those dispositions themselves.

You can even see equivalent traits in temperament of very young infants, which can be summarized in three main factors: Surgency (which tracks positive activity and sensation seeking, similar to Extraversion), Negative Emotionality (which maps onto Neuroticism in adults), and Effortful Control (which corresponds more or less to Conscientiousness). Indeed, you can measure equivalent versions of some of these temperament traits in animals—from cats and dogs to guppies and octopuses.

All of these observations suggest that these designated personality or temperament domains really do tap into underlying biological differences that contribute to differences in how people behave. But crucial questions remain as to whether they reflect really independent biological parameters or dimensions along which people vary. It is possible that these measures are really just statistical artifacts—just a number that we use to describe the correlation between various specific behaviors, rather than valid entities in themselves. Perhaps we are not carving (human) nature at its joints at all. Perhaps it doesn't have joints. Just because we can create some kind of aggregate score does not necessarily mean that it measures a real single thing—it could still reflect multiple, even more basal parameters. Personality data can clearly be sliced and diced in different ways—into 3, 5, 10, 12, 16 factors, or more. There is no strong reason, from the psychological data by themselves, to think that the factors defined by any one of these systems have special biological status.

This is where people have turned to genetics and neuroscience for support. If these factors really do reveal separable aspects of underlying biology, the prediction is that they will correlate with variation in specific sets of genes or specific neural circuits. We will see that this has generally not turned out to be the case. Twin and family studies have clearly shown that personality traits are quite heritable—it is just that the genetic variation contributing to them does not necessarily affect specific kinds of genes or specific neural circuits in the neat way that people had hoped.

SOURCES OF VARIANCE IN PERSONALITY TRAITS

Regardless of the modular status of the variously defined personality traits, it is possible through twin and adoption studies to ask what makes people differ in these traits. Does variation arise innately due to genetic differences or variation in brain development, or does it emerge in response to upbringing and experience?

The findings from such studies are remarkably consistent and extremely well replicated: First, regardless of which scheme is used, almost all such traits are moderately heritable—on the order of 40%–50% of the variation in these traits is attributable to genetic differences between people. This number gets considerably higher (more like 70%) if average values from multiple scorers are used to calculate the personality measures for each person, rather than relying solely on self-report—this tactic clearly increases the validity and reliability of the measures.

Second, there is typically a negligible effect of the shared family environment. Being raised in the same family does *not* make people more similar in personality traits. Conversely, being raised in different families does not make them more different. We have discussed this surprising finding already in chapter 2. Whatever the important effects of nurture and whatever the influences parents have on their children's behavior, these clearly do not reach down to influence the underlying dispositions themselves.

The final conclusion from twin studies is that *some other factor*, in addition to genetic differences, is also contributing to variation in personality between people. Some of this is probably just measurement error—the numbers we assign to these traits are just a bit fuzzy. But, as discussed in chapter 4, a lot of this variation is probably inherently developmental. Differences in the way the program of brain development plays out likely make a significant contribution to people's personalities.

The conclusion from these kinds of studies is that personality traits like the Big Five really are heritable and really do reflect some biological differences. However, they do not speak to the question of whether they reflect *distinct biological modules*—they could emerge as composite measures from statistical analyses without necessarily mapping onto dedicated genes or neural circuits. The only way to figure that out is to try to identify the relevant genes and the circuits involved. That has proved challenging.

SEARCHING FOR GENES AND CIRCUITS

Many studies have been published claiming to have found an association between variation in a specific gene and some personality trait. These began with studies of variants in candidate genes—ones that researchers thought might be involved due to findings from pharmacology, for example. There have been long-standing theories regarding the possible roles of neurochemicals like dopamine or serotonin in traits like Extraversion or Neuroticism. Genes encoding components of these neurochemical pathways (biosynthetic enzymes, receptors, transporters, and others) were thus natural candidates to test for an association with personality traits.

Like all genes, these contain some common variants or SNPs—positions where some people in the population have one version of the DNA sequence and others have another. The test—an association study—is to see whether one of these versions occurs at higher frequency in people who are, say, more extraverted than in people who are less extraverted. The logic of these association tests is very simple: if, at some position in the genome, say, the "A" version of an SNP is significantly more common in extraverts than in introverts, then maybe that variant is having a functional effect that is causing extraversion (see figure 6.1).

Regrettably, the methods employed at the time led to mostly spurious findings. These arose for several reasons: First, the samples used were typically very small (on the order of hundreds of people), which, in retrospect, turned out to be vastly underpowered to detect effects of the size that common variants typically have for other traits. Second, multiple genetic variants, often in multiple genes, were typically tested at once, without correcting the statistics for that fact. If you test one thing and your statistics tell you the likelihood that the frequency difference you observed occurred by chance is only 1 in 20 (a traditional "p-value" cutoff

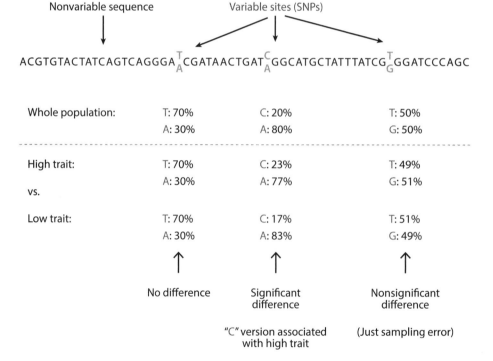

Figure 6.1 Testing single-nucleotide polymorphisms (SNPs) for associations with traits. Some positions in the genome have common variation (SNPs), where two versions persist at some frequency in the population. If one of the versions affects a trait, its frequency should differ between people with high versus low values of that trait.

of 0.05), you might be inclined to think it's real. But if you test 20 different things and 1 of them comes up as statistically significant at that level, well, that's obviously much less convincing—it is exactly what you would expect by chance over 20 trials. Third, these studies usually did not have a separate replication sample, meaning that chance findings were left unchallenged and unreplicated. And, finally, an even more insidious factor was publication bias—studies that find something interesting, like an apparently positive association, are far more likely to be written up and to get published than studies with purely negative results.

All of these factors conspired to generate a literature made up of mostly—perhaps entirely—false positives. We know this is the case because many of the findings failed to replicate in subsequent studies and also because a much more powerful method to carry out these kinds of experiments was subsequently developed—genome-wide association studies (GWAS). GWAS test for association of a trait with genetic variants not just in a single gene but across the entire genome, all at once. Because they test so many variants (usually 500,000 or 1,000,000 SNPs, which effectively capture all the patterns of common variation across the whole genome), the burden of statistical proof for association with any one variant is very high. That means the sample sizes have to be enormous—tens or hundreds of thousands of people. With samples that high, GWAS are powerful enough to detect even very tiny differences in frequency of genetic variants. Also, they use a replication sample to make sure any positive findings are not spurious and report all results for all variants, whether or not they are positive.

GWAS have recently been carried out for Extraversion and Neuroticism in enormous samples and have begun to identify a few associated common variants. Tellingly, none of the previously implicated candidate genes survived this more rigorous testing. Moreover, the results definitively rule out the existence of *any* common variants, in any genes, that have a large or even a modest effect on these traits. The ones that were identified have extremely small statistical effects. Even collectively, the predicted effects of all possible common variants on these traits are quite small.

The other important finding from these GWAS is that the dozen or so genes identified to date do not seem to implicate any particular biochemical pathway or cellular process. It is still early days, perhaps, but so far at least, these traits are not obviously mapping to a recognizable, distinct underlying module at a genetic level.

What about at a neural level? Do the Big Five traits map onto the structure or function of dedicated brain regions or circuits? Well, no, not really. There have been hundreds of studies reporting correlations between personality traits and the size or activity of specific brain regions or circuits, as revealed by magnetic resonance imaging. Unfortunately, these studies have been plagued by the same problems as the candidate gene association studies—small samples, an excessively exploratory approach, lack of replication samples, and substantial publication bias. The result is another literature awash with spurious findings.

However, for both genetic and imaging studies, the lack of consistent positive findings is actually quite informative. It means that assumptions about the functional modularity of the Big Five personality traits—at either the genetic or neural level—are naïve. They clearly do *not* map to variation in a few definable neurochemical pathways, brain regions, or circuits. This suggests that these psychological constructs may not be tagging real unitary "things" after all, at a biological level, but may instead reflect the combined effects of variation in many different cellular and neural systems.

The main problem with the personality traits that have been defined by factor analyses is that they are merely *descriptive* of patterns of behavior. To say individuals like socializing "because they're extraverted" is simply putting a label on that trait—it does not provide any explanatory mechanistic information. But there is a mechanistic context in which we can approach these questions. When we talk about personality traits we are describing different ways that individuals tend to behave in a given situation. This means choosing different actions from a set of possibilities—that is, making a decision. And decision-making can be discussed not just at the output level of patterns of behavior, but also at a deep, mechanistic level in terms of the parameters that the organism takes into account and the kinds of operations involved in integrating them to decide on an appropriate action.

I, ROBOT

Imagine you have to build a robot. It has to be able to make its own way in the world, which requires finding fuel (let's call it food) and avoiding being destroyed by other robots or dangers in its environment, and has to periodically transfer some of its source code from its old, decrepit body to a nice new one (let's call that reproduction and to make it more fun let's have it involve mating with another robot). What would you need to equip your robot with for it to do that job?

Well, it's going to need some kinds of sensors so it can detect things in its environment—potential food or mates or threats. And it'll need some mechanism for locomotion and action. But how can it decide what it should do? How can you program it to make sure it acts in a way that will ensure its survival and reproduction?

The simplest kinds of behavioral programs that could be hardwired into its circuitry would link detection of a particular stimulus to a particular response, equivalent to reflex actions. These might be useful to ensure your robot protects itself from potentially harmful things in its environment—like something dangerously hot, for example. You don't really want your robot spending any time "thinking" about that decision, or worrying about the context, you want the response to be automatic.

But for most behavioral decisions we are going to need something more complex. If your robot is to survive then part of its programming should be to seek out food and eat it when it can. But what if there are potential threats around? In that situation, the robot has to balance the seriousness of the threat with its own drive to eat, which will itself vary depending on how low on fuel it is. It also has to consider opportunity costs—if it spends all its time eating then it won't have any time left to try to find a mate. At any particular moment it will have to weigh up all those factors, based on information about both the environment and its own internal state, and choose to pursue one goal at the expense of the other.

One way to program it to assess situations and decide on the optimal action to take is to get it to assign weights to each of these factors and compute a solution. If the threats or risks are weighted heavily enough then they will outweigh the value assigned to the potential food. But if fuel reserves are dangerously low, the weight given to the food may increase and drive the robot to risk exposure to a threat. Of course, your robot should also be able to learn from experience. It may, for example, have a memory of easily evading a particular threat or of a potential food source actually being toxic, in which case it will assess the particular situation differently.

However, if you want your robot to survive for long enough to learn anything from experience, you're going to have to give it some preset weights for different factors. It will have to start with a certain sensitivity to threats, a certain aversion to risks, a certain drive to mate, a certain level of interest in exploring novel things, and so on. Each of those parameters can be tuned up or down (see figure 6.2). It may take a lot of trial and error to optimize all those different tunings. In fact, because they all interact you will have to find a *combination* of different tunings that maximizes your robot's chance of survival. And there may not be only one solution—there could be many different combinations that all work reasonably well overall, some better in some situations and others better in others.

Now, here's the thing—you may settle on one set, but I might tune my robot differently. Mine might be more sensitive to threats than yours, might weight opportunities for food or for mating more positively, might be more inclined to cooperate with other robots rather than competing, and so on. That would mean that in any given situation my robot would behave differently from yours. Across many situations, these differences would manifest as global tendencies (to be more or less cautious, curious, social, etc.), in ways that would be fairly stable and predictable. In short, our robots would have *personalities*.

I, HUMAN

Something similar seems to underlie human personalities. Assessing different situations and selecting appropriate actions depend on all the same kinds of computations—assigning positive or negative weights to various parameters (threats, opportunities, internal states, short- vs. long-term goals, etc.), and integrating this information to derive an optimal choice of action. We can't get by with lots of fixed stimulus-response reflexes—behavior has to be *organized*. And the main tools that are used to organize behavior along these lines are neuromodulators.

These include molecules like dopamine, serotonin, noradrenaline, acetylcholine, and a huge array of neuropeptides. We encountered them above as hypothetically underlying traits like Extraversion and Neuroticism, but those linkages proved elusive. A much tighter relationship

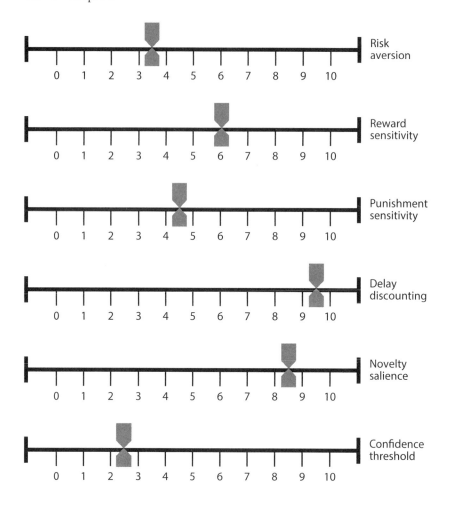

Figure 6.2 Decision-making parameters. A number of distinct decision-making parameters can be independently tuned to different levels across individuals. These may underlie variation in higher-order personality constructs.

may obtain for more basal parameters of decision-making, such as reward sensitivity, risk aversion, or others described below.

In chapter 4 we discussed the basics of neurotransmission—the mechanisms by which neurons communicate with each other. When a neuron is activated it sends an electrical signal down its axon, which triggers release of neurotransmitter molecules at its synapses. These molecules are detected by receptor proteins on the other side of the synapse, in the dendrites of downstream neurons, either exciting those neurons or inhibiting them (depending on the neurotransmitter made by each neuron). But this is only part of the picture. The *strength* of synaptic transmission—how big a signal is sent and how sensitive the downstream neuron is to it—can also be dynamically regulated. This includes changes to the strength of specific local synapses through synaptic plasticity mechanisms, but also more global regulation that reflects motivational states.

Neuromodulators, such as dopamine and serotonin, act broadly to alter the biochemistry of many other neurons over a longer time period, changing how these neurons respond to signals from yet other neurons. They can thus alter what engineers call the *gain* in a circuit, increasing or decreasing sensitivity through one channel or another and shunting information flow on the fly, without having to alter the hardware. They are crucial mediators that convey information about the physiological or motivational state of the organism, including arousal, mood, attention, satiety, and many other parameters.

In decision-making terms, they also help set the parameters that are used in determining what economists call the relative *utility* of any prospective action. Calculating this involves estimating the likelihood of a reward if that action is taken, as well as the subjective value of that reward, versus the likelihood and size of punishment, while also taking into account the amount of time until such a reward or punishment would occur and the opportunity costs of not taking other possible actions. At the same time, the system must assess the quality of the information on which such estimates are based and the degree of associated *uncertainty*, which can itself drive actions aimed at obtaining more information.

The level of signaling of multiple neuromodulators sets the tone of all these computational parameters, thus organizing ongoing behavior and also shaping learning and future behavioral strategies. As discussed in chapter 5, synaptic plasticity can be gated by neuromodulators. This provides a mechanism to regulate learning based on the individual's *subjective* experience—not just how big a reward or punishment was in an objective sense, but how rewarding or punishing it *felt* in a subjective sense. This ties decision-making in with emotions, which may be viewed as heuristic signals that the brain uses to rapidly make approximately optimal decisions with incomplete or ambiguous information.

Over time, this kind of learning leads to the development of habits. In fact, most of the actions we take on a daily basis are habitual—we get up, take a shower, have breakfast, go to work, and so forth. Our brains have learned that in the situation of waking up in the morning, those actions are optimal—we don't have to consciously figure it out again every time. And the same goes for how we act in most situations—the times when we take deliberative decisions are much more rare, and, even in those cases, the range of options that suggest themselves is highly constrained. But in either habitual or deliberative situations, signaling by neuromodulators is a key element in organizing behavior.

Now, you might think all this would mean that the behavior of each individual is infinitely adjustable. If the job of neuromodulators is to dynamically change all those parameters, then surely, by doing just that, they could be used to allow any individual to select any behavioral strategy in any given context. That might be true, except for the fact that the neuromodulator circuits *themselves* are tuned—they work differently in each of us, thus influencing the habitual behavioral strategies we each tend to develop.

VARIATION IN NEUROMODULATORY MACHINERY

In a funny way, some of these ideas were prefigured over 2,000 years ago by the ancient Greek physician Hippocrates' system of the "four humors": blood, yellow bile, black bile, and phlegm. In his scheme, people with different amounts of these would show different patterns of behavior or moods. His successor Galen developed a more complicated scheme also involving four basic elements, which could be combined (or *tempered*) in different ways to yield nine different temperaments. Among these, he labeled ones that had a strong imbalance of elements as sanguine, choleric, phlegmatic, and melancholic—terms we still use colloquially today. In a more modern conception, variation in neuromodulatory signaling pathways may be at least part of what underlies individual personalities.

Neuromodulators like dopamine and serotonin are produced by small populations of neurons in specialized regions in the midbrain (see figure 6.3). These neurons project their axons across large areas of the brain,

DOPAMINE

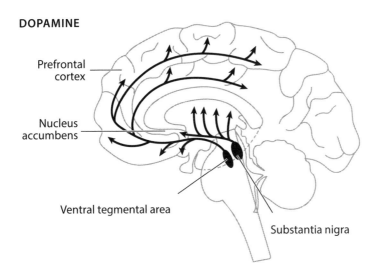

Prefrontal cortex

Nucleus accumbens

Ventral tegmental area

Substantia nigra

SEROTONIN

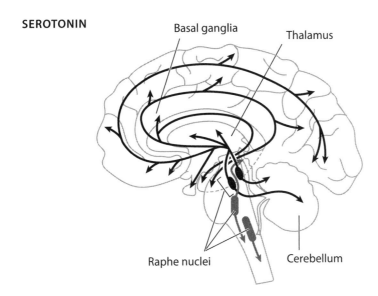

Basal ganglia

Thalamus

Raphe nuclei

Cerebellum

Figure 6.3 Neuromodulatory circuits. Neurons producing neuromodulators such as dopamine and serotonin are found in the midbrain and brainstem. Their axons project throughout the brain (arrows), where they modulate neural transmission in differing ways, via a multitude of selectively expressed receptor proteins.

especially forebrain regions like the cortex and basal ganglia that are involved in decision-making. Neurons in these regions express receptors for these neuromodulators, but these come in many different varieties, which are expressed at different levels in different cell types. This is the general plan, but the precise numbers of dopamine- or serotonin-producing neurons and the extent of their axonal arborization will vary between people. And the biochemistry regulating how much of these neuromodulators is made, released, and responded to will also necessarily vary between people—"necessarily" because there is no way for nature to specify these parameters precisely in all individuals. There are simply too many genes involved to keep them all free of genetic variation and too much noise in the developmental processes for it to run the same way every time.

New neurogenetics technologies are now enabling researchers to dissect the roles of various neuromodulators in decision-making and their contributions to behavioral traits. The important thing about shifting perspective from the merely descriptive to this more mechanistic level is that it enables the development of an explanatory framework of genuinely causal effects from low-level neurobiological parameters to higher-level personality traits. As an illustration, let's consider the trait of impulsivity and the role of serotonin pathways in regulating the computational parameters that feed into it.

IMPULSIVITY

Impulsivity is, broadly, the tendency to act without foresight. But it encompasses multiple facets, such as making a decision without adequate evidence, failing to inhibit an action, prioritizing immediate rewards over long-term ones, or downgrading possible future negative consequences of one's actions. Various aspects of impulsivity feed into Big Five personality traits like Neuroticism (positively, through low self-control), Extraversion (positively, through high sensation seeking), and Conscientiousness (negatively, through low planning and self-discipline). In behavioral terms, one way that impulsivity tends to manifest is in physical aggression—more on that below.

These relationships between psychological constructs can be schematized in many different ways, but the more interesting question for us is what underlying factors make one person more impulsive than another. Impulsivity can be assessed, like other psychological constructs, using various questionnaires. These reveal a trait with fairly typical levels of stability, test-retest reliability, and heritability (around 50%). But these measurements alone are not very illuminating—really they amount to just asking people in many different ways whether or not they tend to behave impulsively.

However, underlying aspects of impulsivity can be very directly measured in experimental tasks of decision-making (see figure 6.4). Crucially, these tasks can be applied to animals just as well as to humans. Monkeys, rats, mice, and even pigeons can be trained to make decisions in laboratory tasks where they have to choose between options, assess evidence, delay gratification, inhibit actions, weigh the probability of future rewards or punishments, change strategies given new information, and so on.

For example, you can train a rat or mouse to choose between stimuli by having it poke its nose into one port or another in an experimental apparatus (people typically use touch screens these days so the animal just pokes a tablet computer). The choice might be to respond to a stimulus that gives an immediate but small reward versus one that

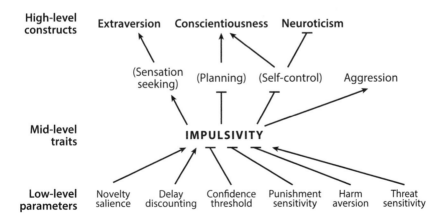

Figure 6.4 Impulsivity. This model shows impulsivity as a mid-level trait, which is affected (positively, *arrows*; or negatively, *T-bars*) by the levels of many lower-level decision-making parameters, and which, in turn, feeds into higher-level personality constructs.

gives a larger reward, which is either delayed or which only occurs with a lower probability. Other tasks require the animal to inhibit premature responses to obtain a reward or to inhibit an already initiated response if a stop signal occurs (like checking a baseball swing on a pitch in the dirt). With a little training, animals, including humans, can get remarkably good at these kinds of tasks and, on average, rapidly approach optimal behavior, maximizing available rewards. But on an individual level, stable differences exist in various aspects of performance. And, most importantly, the underlying neural substrates can be experimentally investigated.

By looking at effects of lesions to various parts of the brain, both in humans and animals, a network of regions has been implicated in these aspects of decision-making. These include the prefrontal cortex (PFC) and orbitofrontal cortex (OFC), at the front of the brain—areas involved generally in what is known as "executive function." These functions, which include assessing evidence, planning, and deciding on actions, are carried out in concert with structures below the cortex known as the basal ganglia and amygdala, among others. The involvement of these regions is supported by evidence from neuroimaging in humans and from neural recordings in animals, which shows which areas are active during these kinds of tasks and even what kind of information is encoded by different sets of neurons.

Crucial to those processes are the neuromodulatory inputs from the brainstem, which convey information that allows the system to compute the utility of current or future states. The effects of various drugs that specifically target different neuromodulator receptors or other signaling components have given invaluable information on the roles of the dopamine, noradrenaline, or serotonin pathways in this system. The functions of serotonin are of particular interest, as they relate particularly to impulsivity.

SEROTONIN AND IMPULSIVITY

In general, low levels of serotonin signaling—both in humans and in animals—are associated with greater impulsivity and increased volatility in behavior, often manifested as hostility and aggression. Serotonin has long been thought to convey signals about punishment and to be involved in negative reinforcement learning. It acts, broadly speaking, in opponency to dopamine, which signals better-than-expected outcomes, or rewards. Recent experiments in mice have revealed some additional subtleties to the functions of serotonin. These have been achieved using the revolutionary technology of optogenetics, which allows millisecond control of activity of specific subsets of neurons.

Optogenetics borrows a trick from nature to allow researchers to turn neurons on or off merely by shining a blue light on them. The trick is a protein, called channelrhodopsin (ChR), which comes from simple, single-celled blue-green algae. It is similar to the opsin proteins that allow photoreceptor cells in our eyes to detect light. In the algae, the ChR protein sits in the membrane of the cell and, when it absorbs a photon of light, it opens a channel that lets electrically charged ions flow into the cell. This is effectively the same mechanism used in neurons to make them fire a signal when they detect neurotransmitter at their synapses. So, if you force neurons to express that ChR protein, then when a blue light is shone on them they become activated. Other related proteins can be used to inactivate neurons instead.

The beautiful thing about this tool is that it can be targeted to very select subsets of cells by hooking up the DNA that encodes the ChR protein to the regulatory elements of some other gene that is expressed just in the cells of interest, and then making transgenic animals that carry this new artificial gene. In the case at hand, the targeted cells include various subsets of serotonin-producing neurons in the midbrain, which project to different regions. Various experiments using this kind of tool have revealed a number of different roles of serotonin during different kinds of tasks.

First, acute activation of the serotonin neurons increases fear and anxiety in the animals. This is consistent with it acting in a circuit that underlies aversive behavior and learning. Fear and anxiety arise as emotional responses to a signal of punishment, which tells the animals not to repeat whatever it was that caused that effect. Serotonin signaling is thus a component of cost assessment and harm aversion and the resultant behavioral inhibition when those signals are high. When serotonin signaling is low, possible negative consequences of actions are downweighted, behavior is less inhibited, and impulsivity is increased.

Separate experiments have also shown a role for serotonin in withholding responses under conditions of incomplete information. Activating the serotonin neurons promotes waiting, while inactivating them leads to an increase in premature, impulsive responses. The same pathways also allow animals to *unlearn* things, in particular to inhibit previously learned responses that are no longer adaptive under current conditions and to drive plasticity to reconfigure the weights assigned to these actions. It is thus a key component of cognitive flexibility—it both suppresses impulsive actions that may be suboptimal or incur negative consequences and prevents animals from persevering in actions that used to pay off but no longer do.

These optogenetic tools provide an unprecedented ability to tease apart the system, to manipulate the underlying computational parameters on the fly, and watch the impact on an animal's behavior. Now we're down into the nuts and bolts, the computational algorithms of decision-making, the circuitry that carries them out, and the signals that organize behavior by tuning those circuits. And, not surprisingly, variation in the genes encoding those components can lead to stable differences in patterns of behavior.

However, the effects of mutations in these genes are not at all simple, and often differ from those observed due to acutely manipulating the encoded proteins in adults, with drugs for example. This is true for three reasons: first, the proteins involved play multiple roles in different brain areas; second, there are often compensatory changes to the levels of other proteins in the system; and, finally, alterations to neuromodulator pathways, especially that of serotonin, also affect the processes of neural development themselves, with knock-on consequences across the brain.

GENETICS OF IMPULSIVITY AND THE SEROTONIN PATHWAY

Serotonin signaling involves a host of specialized proteins, including enzymes that make serotonin—a chemically modified version of the amino acid tryptophan—or that break it down; 14 different receptors, each with discrete biochemical and cellular characteristics; and a protein that hoovers up excess serotonin that has been released at synapses and recycles it, known as the serotonin transporter (see figure 6.5).

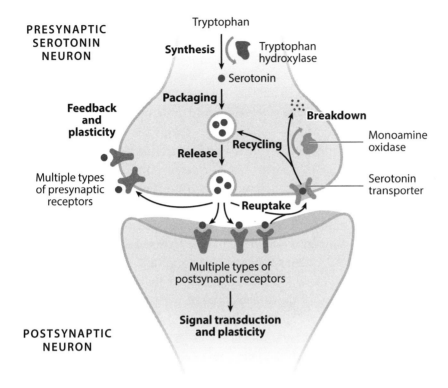

Figure 6.5 Serotonin biochemical pathways. Multiple proteins involved in serotonin synthesis, reuptake, and degradation are shown, along with a diverse array of receptors.

Mutations in many of these genes have been found to affect behavior, often resulting in increased impulsivity and aggression, both in mice and in humans.

In some cases, the results fit with the idea that serotonin mediates behavioral inhibition. In mice, for example, mutations in the gene that encodes tryptophan hydroxylase 2 (*Tph2*), the enzyme that makes serotonin, result in impulsivity and aggression (especially manifested in male mice, which are naturally much more aggressive than female mice). Mutations in a gene encoding one of the serotonin receptors, called *Htr1b*, result in a similar behavioral profile. Conversely, mutations in the gene encoding the serotonin transporter protein—the one that removes excess serotonin from synapses—lead to increased serotonin signaling and result in decreased aggression. So far, so good—all those results fit with a simple model where higher levels of serotonin signaling inhibit impulsive and aggressive behavior.

However, mutations in another serotonin receptor gene, *Htr1a*, have opposite effects, reducing aggression. And mutations in the gene encoding monoamine oxidase-A (*Maoa*), the enzyme that *breaks down* serotonin, result in dramatically increased levels of serotonin in the brain, but also result in impulsivity and aggression—the same profile as in *Tph2* mutant mice, where serotonin levels are depleted.

Collectively, these results highlight the complexity of the biochemical and neural systems involved. There are many, many proteins involved, expressed in different combinations in different cell types, and their functions interact in unexpected ways. In addition, the system *reacts* to changes, such as the absence or reduced activity of one of the components, often by up- or downregulating the levels of other components. To make things more complicated, neuromodulators like serotonin also have direct roles in neural development. Serotonin acts on many developing circuits, affecting gene expression, determining the types of synapses that are made, and regulating activity-dependent plasticity. Altering serotonin signaling during development thus has permanent effects not just on the serotonin circuits themselves but also on other neural circuits in many parts of the brain. This may be a major reason why mutations affecting this pathway, which are obviously present throughout life, including development, have effects that are sometimes hard to reconcile with those caused by acute drug administration or activation of serotonin circuits in adults.

In humans, mutations in some of the same genes also affect impulsivity and aggression. Most notably, a very rare mutation in the *MAOA* gene, which completely knocks out the function of the protein, was discovered in one large Dutch family, where all of the males who carried it showed a syndrome of borderline intellectual disability and dysregulated impulsive aggression. Many had been imprisoned for aberrant sexual behavior, arson, and attempted murder. (Females were less affected as the *MAOA* gene is on the X chromosome, so males only have one copy, while females have a backup.) Similar severe and very rare mutations in *MAOA* have since been discovered in other individuals with comparable behavioral profiles. (More on this gene below.)

Mutations in another serotonin pathway gene have also been found in individuals prone to impulsive aggression, this time in Finland. A mutation in the *HTR2B* gene blocks production of one of the serotonin receptors. In this case, the mutation is more common, occurring in about 2% of the population, probably because of the relatively small founder population of Finland, which caused an arbitrary set of rare mutations to increase in frequency in subsequent generations. The mutation was about twice as common in males who showed a behavioral profile of extreme impulsivity and aggression, especially among violent offenders who had been diagnosed with antisocial personality disorder, borderline personality disorder, or intermittent explosive disorder. In this case, the statistical evidence is not as strong as for mutations in *MAOA* and neither is the effect—clearly, many people in Finland carry this mutation without becoming violent criminals. Whether they show more modestly increased impulsivity on average has not been determined. However, mutations in the *Htr2b* gene in mice also show increased impulsivity, supporting the idea that the mutation in humans really is having an effect.

These findings show that rare mutations with a strong effect on the function of these kinds of proteins can have large effects on behavior, even causing an overt personality disorder. Given that, it is reasonable to ask whether more common genetic variants, which have much more modest effects on protein levels or function, could also affect human personality traits or behavior, but more subtly, contributing to variation across the normal range.

A NOTE ON COMMON VARIANTS AND GENE BY ENVIRONMENT INTERACTIONS

Two genes in particular have been intensively studied in regard to possible effects of common genetic variation, the *MAOA* gene and the gene encoding the serotonin transporter, known as *5HTT*. Both of these genes have a common variant that affects their regulatory regions—the bits of DNA that code for how much of the protein to make. Each of them comes in two versions, one that makes a bit less protein and one that makes a bit more protein. Due to the known functions of these proteins, researchers tested whether either of the two versions (for each of the two genes) was associated with various traits or disorders, like antisocial behavior, neuroticism, anxiety, depression, suicide, and many others. Despite some initial positive results, none of the reported associations has held up.

However, researchers went further, to ask whether these genetic variants might indeed have an effect, but only in people exposed to certain environmental stressors. Two celebrated results emerged. The first is that the low-expression version of the *MAOA* gene is associated with antisocial behavior, but only in those people who experienced childhood maltreatment. The second is that the low-expression version of the *5HTT* gene is associated with depression and suicide attempts, but in a way that depends on exposure to stressful life events. These studies are extremely widely cited, as exemplars of what is called a "gene by environment" (G×E) interaction.

The general idea makes sense—that certain genetic differences would affect vulnerability or resilience to external stressors and only show an overt effect in people exposed to such stressors. Unfortunately, the results themselves don't appear to be reliable. There have been many attempts at replication, some of which have yielded positive results, but others of which showed no or even opposite associations. In general, these kinds of studies have been drastically underpowered, in a statistical sense, which makes generation of spurious findings much more likely. There is also evidence of extensive publication bias in this literature, meaning positive findings are much more likely to be published than negative ones. Moreover, these genes have not shown up as hits in the massive GWAS that have subsequently been carried out for personality traits or disorders like depression. Even if their effects were seen in only a subset of carriers, these studies should have been large enough to detect them.

THE CENTRAL ROLE OF DEVELOPMENT

Overall, the genetic findings show that severe mutations in genes encoding components of neuromodulatory pathways *can* affect the kinds of decision-making parameters that seem to underlie personality traits. However, the mutations identified to date—the ones with large effects—are very rare. On the other hand, common genetic variants don't seem to have much of an effect. That means that the genetic architecture of personality traits is most likely dominated by rare variants with smaller effects. There is also some evidence suggesting that these effects will interact with each other in nonadditive ways, so that the specific *combination* of variants in each individual is what matters most. Unfortunately, both of these factors will make it much harder to identify the specific genetic variants influencing personality traits in any individual. That doesn't make these traits any less heritable; it just means their genetics are complicated.

Having gone from superficial psychological traits down to the level of decision-making circuits, there is one more shift in perspective we have to make. The discussion above has centered on genetic variation in the components of the systems used to organize behavior. Those are the proteins and circuits that we could say *do that job*. But that job is also indirectly dependent on the functions of thousands of other proteins, especially those involved in development of these circuits. Mutations in any of those genes might lead to variation in how these circuits form, and indirectly affect the kinds of decision-making computations we've been talking about. In fact, the majority of mutations that affect these processes likely fall in genes that we would *not* identify as directly involved in the decision-making machinery itself.

There are dozens of examples of mutations in mice in neurodevelopmental genes that indirectly affect the neural systems of decision-making and lead to phenotypes like increased impulsivity and aggression (or affect activity levels, anxiety, sociability, threat sensitivity, risk aversion, etc.). We've studied several such cases in my own lab where mutations that primarily affect cell migration, axon guidance, or synapse formation indirectly lead to circuit changes that manifest with these kinds of consistent behavioral effects.

The same is true in humans—mutations that result in neurodevelopmental disorders are often associated with particular behavioral or personality profiles. These kinds of mutations typically affect many systems, but that doesn't make their contribution to variation in the subcomponents of personality traits any less important. While we might like to try and pigeonhole different genes into specific roles, we're really fooling ourselves. Nature is under no obligation to make things simple and easy for us to understand.

To summarize, the things we recognize and classify as personality traits may be built up from variation in a large number of more basal decision-making parameters. Variation in neuromodulatory systems may underlie the differential tuning of these parameters across individuals. Some of that variation may be due to genetic variants in genes encoding the biochemical components of those neuromodulatory pathways themselves. However, more of the variation is likely far less specific, indirectly affecting those systems, predominantly through effects on neural development.

This brings us back to a central theme of this book—that variation in how brain circuits *develop* makes a major contribution to our psychological traits. Crucially, that variation can arise from genetic differences, but also from the processes of development themselves. Stochastic developmental variation will play a large role in determining how the effects of genetic variation are played out in each of us, and will also contribute to innate differences in temperament. And, as discussed in the previous chapter, the ongoing self-organizing nature of postnatal brain development will tend to reinforce innate differences by affecting the experiences we have and the ways we respond to them. We really are born different and, in many ways, we get more so over time.

CHAPTER 7

DO YOU SEE WHAT I SEE?

Do we all see things the same way? This is a question that has occupied philosophers for millennia, mainly because it is almost impossible to answer. At least, it is effectively impossible—maybe even impossible in principle—to show that two people are having the *same* subjective perceptual experience. When I see a red apple, is the quality of my experience the same as yours? How could we tell? We may be able to show in some way that the *content* of our experience is more or less equivalent (we can both report seeing a red apple; we may even have similar brain activation patterns), but perception is such an intrinsically subjective and essentially private process that the *quality* of our experiences seems almost impenetrable to science. Does the redness feel the same to me as it does to you?

Well, while we may never be able to demonstrate that two people are having the same subjective perceptual experience, there are certainly many cases in which it is clear that they are having *different* experiences. Perception is not just a passive process of detecting and parsing incoming stimuli. It is a highly active process, or really a collection of many separable processes, that together allow us to generate an idea of what is out in the world. Those processes rely on incredibly sophisticated neural circuitry, which is assembled based on instructions encoded in the genome. Genetic variation in those instructions can have dramatic effects on how perceptual circuits are organized, leading to substantial variation in perceptual experiences. This affects not just how we all subjectively experience various aspects of the world but also, at a very fundamental level, how we think about them.

PERCEPTION AS ACTIVE INFERENCE

From an ecological perspective, the point of perception is to allow an organism to figure out what is out in the world around it—where are there objects, what are they, which ones are moving, which ones can I eat, which ones will eat me? Our sensory systems can detect only certain kinds of stimuli—photons of light, air vibrations, heat, pressure from surfaces we are in contact with, chemicals in our environment. From that information our higher-order perceptual systems have to infer what it is, out in the world, that is the source or the cause of those stimuli.

That is a tricky and difficult job. It is easy to predict, for example, what pattern of stimulation a certain object will make on the retina of someone looking at it—it is far more difficult to infer from any given pattern of stimulation what the object causing it is. The information we receive is ambiguous, the detectors are imperfect, the signals are noisy, and there are usually multiple possible solutions to this "inverse problem." Is it small or far away? Is it vertically oriented or slanted toward us? Is it a bright object in dim light or a dark object in bright light?

The first job is to parse that incoming information to try to extract as much meaning from it as possible. For example, in the visual system, layers of cells in the retina perform sophisticated computations on the incoming information by comparing inputs across different cells (see figure 7.1A). Light is detected initially by photoreceptor cells called rods and cones, which express special proteins called opsins that absorb incoming photons and activate a biochemical signal inside the cell, which ultimately determines its electrical activity. Rods are more sensitive than cones and are excited by photons of many wavelengths. Cones are more selective—they come in three varieties that respond preferentially to light of different wavelengths.

The activity of the photoreceptors is monitored by bipolar cells, which can be either activated (turned "ON") or inhibited (turned "OFF") by the photoreceptor signals. Each bipolar cell responds to either a rod or a cone cell, but its activity is also influenced by the state of the neighboring photoreceptors, through the action of another type of cell, called horizontal cells. In turn, bipolar cells synapse onto retinal ganglion cells, which are the output cells of the retina—the ones that carry a signal

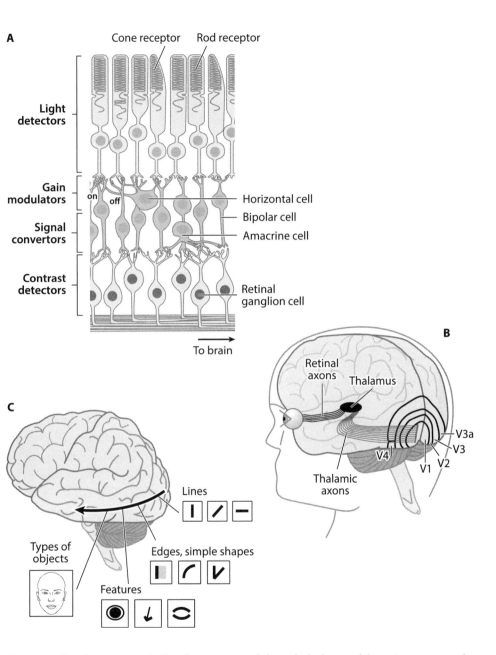

Figure 7.1 Visual processing. **A**. Signals are processed through the layers of the retina to extract the most meaningful elements of the visual scene. **B**. Retinal axons project to the thalamus, and thalamic axons project in turn to the primary visual cortex (V1), and so on through 30 different visual areas (V2, V3, V3a, V4, and many others). **C**. Higher-order visual properties are extracted through this hierarchy, from lines to simple shapes to features of objects, to specific types of objects, such as faces.

through the optic nerve to the brain. Retinal ganglion cells integrate information from multiple neighboring ON and OFF bipolar cells at once and build up a more detailed picture of the visual stimulus. In particular, they are contrast detectors—they will be most active when there is a *difference* in activity between neighboring bipolar cells. That means they pay most attention to the edges of objects and are less interested in solid surfaces.

Something similar happens with signals from the cone cells. These each express one opsin gene, which is sensitive to wavelengths of light that we see as red, green, or blue. (The choice of which of the opsin genes to turn on is made at random by each cone cell, another example of development not being deterministic.) Inputs to cones expressing the red and the green opsins are compared with each other in one channel of bipolar and ganglion cells. In turn, inputs in the red/green channel are compared with the blue channel, giving a blue/yellow contrast. Together, these comparisons give us our rich experience of millions of shades and hues.

There are many more subtypes of retinal ganglion cells that are performing different kinds of computations, based on the architecture and logic of their inputs. Some are responsive to movement, for example, in one direction or another. Some are more responsive to flickering than sustained light. Some pass a fast signal that is low resolution, while others pass a slightly slower signal that carries more detail. Others send information to mediate nonconscious visual responses, controlling things like pupillary reflexes or circadian rhythms. Thus, even within the very first station of visual processing—the eye itself—the circuitry is extremely complex and specialized for extracting specific features of the visual signal and passing that information on, in multiple parallel streams, to the brain.

The signals that actually mediate visual experience are transmitted first to the thalamus in the very middle of the brain, and then on to the primary visual cortex, at the back of the head (see figure 7.1B). The nerve projections that carry this information have a very important property—they make connections to their target cells in a way that maintains nearest neighbor relationships. Cells that are next to each other in the retina project to cells that are next to each other in the thalamus, and so on to the visual cortex. Since the visual world is mapped across the surface of the retina by the lens of the eye, that means it is also mapped across the surface of the primary visual cortex.

But it doesn't stop there. There are at least 30 more areas of cortex devoted to vision, which are arranged in a roughly hierarchical fashion. Each stage integrates information from the stage below in order to extract more and more complex visual features. Primary visual cortex cells integrate information from the thalamus and retina (mainly signifying dots of high contrast) and thereby detect lines of various orientations. Higher-order areas respond to simple shapes or curves, and then to objects, with some areas specifically interested in color or motion or other visual features. Eventually we get to areas that are really specialized for *types* of objects—like faces or letters or tools or houses (see figure 7.1C).

The pattern of activity across all these visual areas thus reflects the properties of the objects out in the visual world. If we recorded these patterns with electrodes or by neuroimaging, we would have a good chance of "decoding" what the person is seeing by looking at those patterns. (Researchers have even made progress at decoding the contents of dreams with this approach.) But here's the thing—there is no one in your brain looking at the patterns. There's no little person in there staring at these neural projections. None of the individual neurons "sees" anything. The retina doesn't see anything, the thalamus doesn't see anything, nor do any of the many visual areas of the cortex. Just parsing and passing those patterns does not constitute *seeing*.

Instead, seeing (or perception more generally) occurs through the act of inferring what it is that is causing the sensory stimuli. That is thought to happen when our brains compare the incoming sensory information to an internal "model" we have (or that our brains have, at least) of the current state of the world around us (see figure 7.2). The idea is that the brain then adjusts the model to accommodate those signals. The brain is, in essence, predicting the state of the world and deriving a measure of the error of that prediction by comparing it with the sensory data. It then tries to reduce that error to zero by updating the model. Somehow, that process of adjusting the model is thought to underlie conscious subjective perception. (The "somehow" in that statement massively underplays the degree of mystery surrounding how that actually happens!)

This comparison requires information to flow in both directions— not just bottom-up, from the sensory periphery to higher and higher

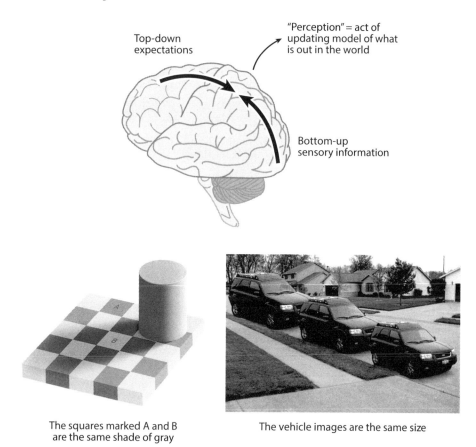

The squares marked A and B
are the same shade of gray

The vehicle images are the same size

Figure 7.2 Perception as inference. Perception involves an updating of an internal model of the world, based on a comparison of incoming sensory information with top-down expectations. Bottom panels show two illusions that illustrate the strength of top-down effects. (Left panel reprinted from Wikipedia contributors, "Checker Shadow Illusion," *Wikipedia, The Free Encyclopedia*, March 9, 2018, https://en.wikipedia.org/w/index.php?title=Checker_shadow _illusion&oldid=829646768; right panel reprinted from "Logic Optical Illusions," *Genius Puzzles*, n. d., https://gpuzzles.com/optical-illusions/logic/.)

regions of the cortex, but also top-down, to carry the information about the current model. Indeed, most of the inputs to the primary visual cortex are feedback connections from the higher visual areas, rather than feed-forward connections carrying sensory information from the thalamus and retina. The influence of those top-down connections can be revealed in optical illusions, especially in conditions where our knowledge or expectations of the way the world *should be* changes our perception, overriding lower-level signals and altering what we actually see.

A couple of examples are shown in figure 7.2. In the panel with the checkerboard, the two squares labeled "A" and "B" appear to be different shades of gray. In fact, they are exactly the same. Your brain interprets one as brighter than the other because one appears to be in the shade—for it to be giving off the same light intensity, as detected by your retina, your brain figures out that the object there *must actually be brighter*, which is what you see. Similarly, in the picture of the SUVs, the vehicle on the right looks smaller than the one on the left, though the images are the same size. The perceived perspective of the road makes one seem farther away than the other. If it nevertheless takes up as much room on your retina, your brain infers that it *must actually be bigger*. These examples dramatically illustrate that the job of perception is to build a representation or best guess of what is out in the world, not just to passively propagate sensory signals.

All of that effort of perception is toward a purpose. What an organism needs to know is not just what something is, but what *meaning* it has. What can I do with it? Is it dangerous? Can I pick it up? Can I eat it? Is it tasty? Can I mate with it? Is it going to try to eat me? This leads to something really interesting, because that meaning is different for different organisms. There are different subsets of things in the world that are salient for different kinds of animals—some that a particular species must pay particular attention to and others that it can afford to ignore. Each species is highly adapted to sense the things that matter to it, while often being completely oblivious to things that are irrelevant to its survival (or, more to the point, the survival of its genes). This means, in a very real way, that different species inhabit different environments, at least subjectively, which is, after all, the only experience we have.

THE *UMWELT*

This idea of the perceptual world that each species inhabits was dubbed the *Umwelt* by the German biologist Jakob von Uexküll (the word translates literally as environment, but here means something closer to sensory milieu). He described each species as living in its own bubble, where it perceives only a fraction of the world around it—only the elements that have meaning for it.

For example, bees respond to patterns on the surface of flowers in the ultraviolet range of wavelengths, which humans are completely unaware of. Many birds can see into the ultraviolet range too—hawks use this to track the trails of their prey. Some snakes can detect light in the infrared range, using a specialized organ on their nose, which enables them to sense warm objects from up to a meter away. Bats and rodents can detect sounds at frequencies far beyond human hearing. Dogs can detect a much greater range of odors with far greater sensitivity than we can. These differences even extend into totally distinct senses, which humans don't have at all. For example, platypuses and electric eels detect electrical fields given off by their prey. Octopuses can sense the polarization of light (the plane of oscillation of the light waves), which allows them to "see" otherwise transparent prey. Many other species—such as turtles, bees, and some birds—can even sense the earth's magnetic field.

There are thus vast differences in each species' sensorium—the range of things it can detect at all. But there are also differences in the quality of these sensations, especially in resolution. This is most obvious, perhaps, in color vision. The number of separate color channels that an organism can compare depends on how many opsin genes it has. Many species of mammals have only two, which means the range of colors they can discriminate is much smaller than that for humans. Some others, though, have more—as many as 15 different opsin genes have been found in one species of butterfly, which may enable finer discrimination of light of different wavelengths. Differences in discriminatory resolution between species are also widespread in hearing, smell, touch, and other senses.

Amazingly, species also differ in how *fast* they perceive the world. Our vision may seem like a seamless stream of images but really we see the world in about 60 frames per second. This can be shown by testing what rate of flickering we can detect in a light. If it's faster than 60 Hz (i.e., 60 cycles per second), then we can't detect it—the light just looks constant. But we can detect the flicker of lights at lower frequencies. This frame rate differs hugely across species, as shown by my colleague Andrew Jackson and others—sharks see at around 30 Hz (only half as fast as we do), but dogs see twice as fast as us, at 120 Hz (which may explain why they're so good at catching balls). But the winners are insects, which see up to seven times faster than humans! That makes sense, in that they also move very fast, so in order for the world not to become a blur around them, they have to process more information per second. It may also explain why it's so hard to swat a fly—like Keanu Reeves dodging bullets in *The Matrix*,[1] they may see our hand coming in what seems like slow motion to them.

Beyond the machinery for detection and resolution, there are also crucial differences in what these various stimuli *mean* for different species. This is particularly obvious for various smells and tastes. Different chemicals are innately attractive or repulsive to different species—they are not just detected, they smell *good* or *bad*. For example, mammalian feces smell bad to humans, but, apparently, smell wonderful to flies.

All of these differences in perception between species are innate and hardwired. They reflect differences in biochemistry and neural circuitry that are themselves based on differences in the genetic programs of development. It takes a lot of genetic instructions to wire up all those specialized systems just so. And, as with other traits, the geneticist's version of Murphy's Law applies—anything that can vary will. This means that genetic variation *within species* can also lead to differences in perception between individuals, including in humans.

PERCEPTUAL DIFFERENCES IN HUMANS

There are many examples of differences in perception in humans, the most obvious of which start with simple differences in the sensory apparatus itself. Sensation begins with specialized receptor proteins, expressed by specialized cells, embedded in specialized circuitry. Mutations affecting any of those aspects can result in a change in or even complete loss of sensation.

There are over 400 distinct genetic syndromes that can cause congenital deafness or hearing impairment, for example, which collectively

---

[1] *The Matrix*, directed by the Wachowski brothers (United States: Village Roadshow Pictures, Groucho II Film Partnership, and Silver Pictures, March 31, 1999).

affect about 1% of the population. Hearing depends on specialized hair cells in the cochlea of the inner ear, which sense air vibrations and transmit that signal to the brain. Movement of these cells is detected by a complex molecular apparatus of proteins inside each cell and other proteins forming links between cells. Some mutations that cause deafness specifically affect these proteins, while others have more indirect effects on the development or survival of the hair cells or the formation of the auditory nerve.

Similarly, mutations in genes encoding receptors for various chemicals can lead to deficits or differences in the detection of specific odors or the ability to taste specific compounds. Odorants are detected by olfactory neurons in the nose, which each express only a single gene from a repertoire of about 1,000 different receptor protein genes. These proteins sit in the membrane of the cell and each one is evolved to very specifically bind a different chemical compound. When it does, it sends a signal to the brain that that compound has been detected, which leads to the subjective experience of a smell. Mutations in these genes are common in humans and we all therefore differ in our ability to detect various odorants. In fact, many of those 1,000 genes are nonfunctional in humans, as we don't rely on our sense of smell very much, compared with other animals like dogs or elephants, which have very large functional repertoires of odorant receptor genes.

The same is true for taste—this also relies on a set of specific receptor proteins that detect sweetness (sugar), sourness (literally protons from acidic substances), saltiness (sodium), umami (the rich, round taste of monosodium glutamate), fat (a recently discovered separate taste), and bitterness. Many different compounds taste bitter to us and detecting them all relies on many different genes. Each of these encodes a different receptor protein, but when they are activated they all send the same simple signal to the brain—that tastes bad! It might poison us! We should not eat that! Mutations in taste receptor genes are actually quite common in humans, particularly in the 2 sour receptor genes or any of the 25 or so known bitter receptor genes. A famous example affects the ability to detect a compound called phenylthiocarbamide. About 16% of the population cannot detect that due to mutations in a specific bitter receptor gene. This variation affects whether people taste things like cucumber or brussels sprouts as bitter and is even associated with smoking—people who cannot taste that bitter compound, which is a component of cigarette smoke, are more likely to be smokers.

This kind of variation extends to our mechanical senses too. We have multiple types of neurons that have sensory endings in our skin, which are specialized to detect touch, vibration, pressure, heat, cold, itch, and pain. Each of these relies on different receptor proteins to specifically detect these stimuli, and mutations in the genes encoding those receptors can thus selectively impair one type of sensation but not the others. There are also discrete sets of genes that control the differentiation, wiring, and survival of these different types of neurons, mutations in which can also affect mechanosensation. For example, congenital insensitivity to pain can be caused by mutations in many different genes, with diverse functions, some affecting pain sensation directly, others affecting the survival of multiple subtypes of neurons. Being insensitive to pain may sound like a good thing, but it is actually a debilitating medical condition, as patients often suffer injuries without knowing it, even hurting themselves inadvertently; for example, by rubbing or scratching their eyes too hard or biting off bits of their tongue. There's a very good reason we feel pain.

We sense things inside our own bodies, too, especially the positions of our joints and muscles in space and the tension they are under. This sense, called proprioception, is essential for coordinated movement, and relies on another set of specialized nerve fibers that innervate the joints and muscles. These also express specific mechanoreceptor proteins, and mutations in the genes that encode those receptors cause people to be highly uncoordinated, with unusual posture and muscle tone, even leading to progressive scoliosis (lateral curvature of the spine).

In vision, there are many genetic conditions that cause blindness, some from birth, but more due to a progressive degeneration of the photoreceptor cells. And there are hereditary conditions like myopia or astigmatism that affect the shape of the eyeball or the properties of the lens and impair the ability to focus visual images on the retina. These affect the spatial resolution of vision, though they are commonly corrected with glasses or contact lenses.

Color blindness is one of the most common perceptual conditions, affecting around 8% of males. The genes encoding the red- and green-sensitive opsin proteins are located on the X chromosome (see figure 7.3). A mutation in one or other of those genes will leave males unable to

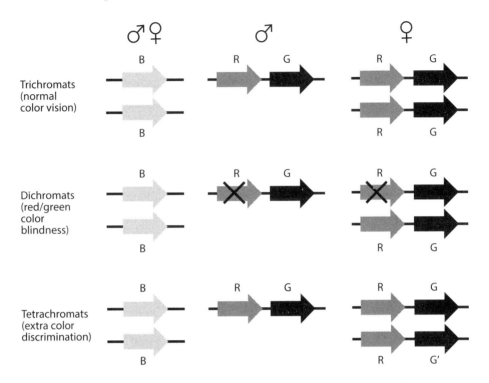

Figure 7.3 Color vision. Males and females carry two copies of the gene encoding the opsin sensitive to blue light (B). The two genes encoding the red (R) and green (G) opsins are on the X chromosome, so males only carry a single copy of each. Loss of function of one of the R or G genes leads to dichromacy (color blindness) in males. Altered function (G′) can lead to tetrachromacy in females.

make the comparison between those two channels, meaning they cannot discriminate between red and green (and cannot use the relative strengths of those inputs to help discriminate across the whole spectrum). This is in fact the ancestral situation present in Old World monkeys—they have only one opsin gene on the X chromosome and thus have dichromatic rather than trichromatic vision. That single opsin gene was duplicated in the lineage leading to apes and humans, and the two copies came to encode slightly different opsin proteins, which absorb red or green light preferentially. Because females have two copies of the X chromosome, they usually have a functional backup copy and so are unaffected carriers of the mutation.

There are also more subtle mutations in the opsin genes that don't completely block function (as in color blindness)—instead, they subtly alter the wavelength of light that the protein absorbs. In males this will mean that the peak wavelength of one of the color channels is shifted up or down, resulting in a different comparison with other channels and presumably altered subjective color perception (though this is difficult to demonstrate). In females, however, there may be a more striking effect, due to the fact that each cell randomly inactivates one of the copies of the X chromosome.

Each cone cell in the retina "chooses" to express either the red or the green opsin gene, at random. However, if a female has a mutation in one of the opsin genes on the X chromosome that alters the wavelength of light that the protein absorbs, this will mean that individual cones will express either that altered version or the normal one, depending on which X chromosome they inactivate. In animals with only a single red/green opsin gene on the X chromosome, this leads to a third functional channel that can be used for color discrimination, meaning that females carrying such mutations are trichromats. In humans, it means that some females are tetrachromats, with four functional channels.

Most people (trichromats) will say they can discriminate about 7 broad bands of color in the visible spectrum, in a rainbow for example. People with red-green color blindness (dichromats) typically report 5. While the research into it is still pretty preliminary, it seems that women who are tetrachromats can distinguish about 10 bands. This calls to mind the consistent observation that women tend to use many more words for colors than men do. Where men may say "yellow," "green," or "blue," women may say "mustard," "dark sage," or "teal." However, it is estimated that only 2% of women have four functional cone types, so the idea that this difference in vocabulary is actually caused by this low level of tetrachromacy seems unlikely.

All of those are pretty extreme examples, where a receptor protein has been rendered completely functionless, or where the sensory cells or their connections are drastically affected. But you can imagine that other, less drastic genetic variation could lead to more subtle variation in the biochemical characteristics of the receptor proteins or in the number of sensory cells or density of their connections, which might in turn affect how sensitive individuals are to sounds, or pain, or cold, or itch, or any of those discrete sensations. Indeed, in vision there is clear evidence that heritable variation in circuitry can affect both spatial and temporal resolution across the normal range.

Spatial resolution is correlated with variation in the size of the primary visual cortex (V1). This region, at the very back of the cortex, shows up to a threefold difference in surface area between people. These differences are largely genetic in origin, driven both by genetic variants affecting general brain size and by ones specifically affecting visual cortical areas. Professor Geraint Rees and colleagues at University College London have shown that a larger visual cortex effectively gives more room to spread out the signal from the retina and correlates with greater spatial resolution and ability to detect fine details. There is a trade-off, however, as people with a larger visual cortex spatially integrate signals from smaller regions of the retina. This makes sense—if integration occurs over a certain-sized cluster of neurons in V1, then this will represent a smaller area of the retina (and, thus, the visual world) if V1 is larger. That spatial integration is important for extracting information about the relationships between neighboring objects in the visual scene. Rees and colleagues have shown, using some clever optical illusions, that the difference in surface area of V1 correlates with differences in *subjective* experience of these relationships.

Just as with many different species, humans also differ in how fast they perceive the world. Our ability to detect a flicker in a stimulus goes up to 50 or 60 Hz (flashes per second)—anything faster than that looks continuous. But we also have higher-order processes of visual cognition performing more complex computations that are a good bit slower than that—more on the order of 10 Hz. That means anything happening faster than 10 times a second (once every 100 milliseconds) gets blurred and is impossible for us to distinguish. However, that perceptual "frame rate" (really an integration window) varies between people.

This can be tested by, for example, showing people a dot flashing twice on a screen. If the flashes occur within a single integration window, they appear as only one, but if they are farther apart in time they can be distinguished as two. The length of that window can be measured and it differs between people. Remarkably, it is strongly correlated with the peak frequency of a particular brain oscillation—the so-called alpha wave of the EEG. This oscillation is strongly detected over the visual cortex and is around 10 Hz on average, but it varies considerably, from 8 to 13 Hz. It reflects a synchronized oscillation in electrical excitability of large populations of neurons—they are highly excitable at the peak of the oscillation and much less so at the trough. People with a higher-frequency alpha oscillation have a shorter integration window and are therefore able to distinguish two flashes closer together. A similar correlation is seen for integration of a visual and an auditory stimulus. As the peak alpha frequency is at least 50% heritable, this is another example of how continuous variation in how the brain is wired (across the normal range) alters subjective perception, in this case causing some people to see the world faster than others.

FROM PERCEPTS TO CONCEPTS

Perception is a skill—one that we get better at over the first few years of our lives and beyond. Through experience, we learn to not just detect objects in the world but to categorize them into *types* of things—living, nonliving, animals, people, dogs, stones, buildings, tools, toys, food, and so on. When we perceive a thing we do more than just process its sensory attributes—we compare them to our memories of past experiences to categorize and recognize it. That act of recognition is made based on whatever subset of sensory attributes the object is currently presenting to us, but it also involves mentally accessing its other attributes from our memories of that object, or type of object.

It turns out that's not so straightforward. It relies on extended neural circuitry linking perceptual regions with areas involved in memory and conscious awareness. There are many examples of perceptual conditions that specifically affect these processes of recognition and access to our knowledge of the attributes of an object. Probably the best known of these is face blindness (or prosopagnosia, from *prosop*—face, and *agnosia*—lack of knowledge of). It was highlighted in Oliver Sacks's celebrated book *The Man Who Mistook His Wife for a Hat*. Sacks described the eponymous patient, a Dr. P., who was referred to Sacks's neurology clinic due to increasing trouble in recognizing people. This was so severe that at one point he literally did mistake his wife's face for his hat and tried to lift it up to put it on. Dr. P.'s deficit turned out to be more widespread, as he also had trouble in recognizing objects, though he could still see perfectly well.

The cause of this particular case remained mysterious, but it was clearly acquired somehow in that the subject had not previously had such difficulties. There are only a handful of case reports of acquired prosopagnosia, mostly due to head injuries, and for a long time it was thought to be extremely rare. However, it is now clear that as many as 2% of the population are born with a specific deficit in face recognition. The condition is very strongly genetic and seems to run in families in a "Mendelian" fashion, which means it looks like it is caused by a single mutation in each family, rather than many genetic variants acting in combination.

We don't yet know what the responsible genes are, but the study of people with congenital prosopagnosia has led to a hypothesis as to what is going wrong on a neural level. As you go higher and higher in the hierarchy of cortical areas devoted to vision, you get to areas that are very selectively responsive to particular types of visual objects. One of those is particularly interested in faces, so much so that it has been named the "fusiform face area" (being located in a part of the brain called the fusiform gyrus). In neuroimaging experiments, that area lights up like crazy when you see a face, but not so much to other stimuli. Of course, it's not sitting there doing anything by itself—it is actually part of an extended circuit of regions, which are all also highly responsive to faces (see figure 7.4).

Interestingly, in people with face blindness, the face area (or network of areas) seems to be normally responsive to faces—those areas of the brain light up just as in other people. And the activity of those areas is even sensitive to whether or not the face has been seen before—at some level, the brain is performing the initial steps of facial recognition perfectly well. The difference is that the face network fails to communicate this signal to the frontal parts of the brain. Ordinarily, there is a later signal in the frontal lobes that correlates with conscious perception of a face. In prosopagnosics, that signal is absent or much reduced. This correlates with observations of fewer nerve connections running between these areas than in control subjects. Presumably, whichever gene is disrupted (most likely different ones in different families), it is either directly involved in or indirectly required for formation of those connections.

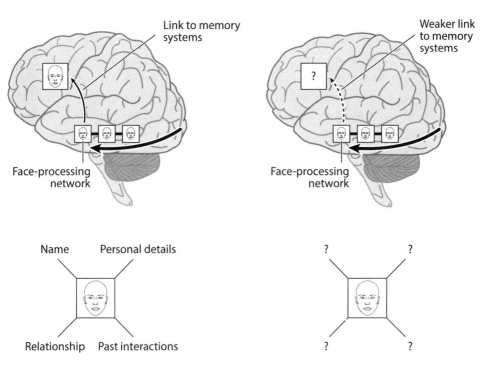

Figure 7.4 Face perception. *Left*: faces are processed through a series of specialized visual areas (*big arrow*). Familiar faces activate frontal memory systems (*small arrow*), allowing recognition and recall of knowledge of that person. *Right*: in face blindness, the specialized visual areas respond to faces, but frontal areas do not, which may underlie failure to recall the person's identity.

The upshot is that people with face blindness can see faces, they can even read emotional signals in faces, they just can't *recognize* the face because they can't link it to stored memory of the person involved. They can't use the face stimulus itself to draw up all that other information—the person's name, the nature of their relationship to the person, or indeed any other details about that person at all. The condition can be so severe that sufferers may be unable to recognize close family members or even pick out pictures of their own face from a lineup! This is obviously severely debilitating in a species as social as ours, where facial differences represent the primary cues for recognition. To compensate, many prosopagnosics develop alternative strategies to recognize people—by their voice, haircut, clothes, jewelry, gait, or particularly distinctive specific facial features (in the absence of accessing an overall impression).

Tone deafness is another example that seems to have a very similar underpinning. It is better labeled "tune deafness," really (or congenital amusia). People with this condition—about 3% of the population—have normal hearing and can usually discriminate between musical notes perfectly well; that is, if they are presented with one note after another they can tell if they are the same or different. Where they struggle is in making those judgments based on knowledge of the structure of a melody or tune. They may be completely unable to detect a note of the wrong pitch in the context of a familiar melody; though, as with face blindness, it seems their brains can do this quite well. EEG recordings can detect a rapid and reliable response to a misplaced note, both in control subjects and congenital amusics. But, as in face perception, there is a second signal that occurs slightly later, in frontal regions, which correlates with conscious awareness of that wrong note—this is seen in control subjects but not in the amusic subjects. Once again, the initial stimulus is processed normally but the conscious subjective percept is very different. This condition is usually also characterized by an inability to recognize familiar tunes from the melody alone (without lyrics) and, not surprisingly, difficulties in singing in tune.

As with face blindness, congenital amusia has a strong genetic underpinning, with rates in siblings of affected people over tenfold higher than in the general population. Both of these conditions represent a severe deficit in a specific aspect of perception. But both those faculties—recognizing faces or detecting incorrect notes in melodies—also vary continuously and measurably across the normal range (independently of other traits, like IQ). Twin and family studies have shown that this variation too is highly genetic. This suggests a situation similar to that for the genetics of intelligence—there is a statistically normal distribution of ability, most likely caused by a combination of many genetic variants in each person, and there is also a discrete category of people with particular impairment, most likely due to single mutations (as for intellectual disability).

There are many other kinds of agnosias that are similarly selective, but for different kinds of stimuli. Dyslexia is another common one, where people struggle with linking either the visual shapes of letters to the sounds they make, or the visual shapes of words to the meanings or concepts of the words. It's not that this can't be done, just that it remains effortful for people with dyslexia, while it becomes highly automatic for others. The heritability of dyslexia is about 50%; though, again, the genetic variants that contribute to it remain largely elusive. As with most disorders affecting neural development, what is most likely inherited is a certain probability to develop the condition, with the actual outcome also affected by stochastic developmental variation.

Color agnosia is a far less studied condition, known primarily from acquired cases of injury to specific "color knowledge" areas in the visual cortex. But there are some reports of apparently congenital cases, even ones that run in families. People with this condition are not color-blind—they can see color perfectly well and can discriminate between hues absolutely normally. What they can't do is link the color stimuli to concepts. They can't, for example, name specific colors, though they can see the difference between them—they have no concept labeled "red" or "yellow." And they can't incorporate color information or memory into the *schemas* or concepts of the attributes of objects—they not only couldn't say that strawberries are red, they also would not be surprised to see a blue one.

Collectively, agnosias are characterized by a difficulty in connecting percepts to concepts and incorporating all the attributes of an object into a schema. This may be due to reduced connectivity between areas that process different attributes or between perceptual and frontal areas that connect with memory and conscious awareness. Remarkably, there is also a condition that seems to be the opposite—where *additional* perceptual attributes are generated internally by the brain itself and incorporated into the schemas of various objects. This condition, called synesthesia, may be due to hyperconnectivity between areas of the brain that are normally not talking to each other.

SYNESTHESIA

People with synesthesia (which means a "mixing of the senses") experience additional percepts in response to stimulation of one sense or other. They may, for example, see flashes of light or color when they hear sounds—either general sounds or specific types, like music. But there are many different types—over 80 recognized so far, in fact, which involve different combinations of inducing stimuli (sounds, tastes, words, numbers, music, people) and different concurrent percepts (colors, sounds, tastes, odors, touch, etc.).

Some synesthetes taste words, for example—hearing or even thinking of a word can produce a strong and consistent taste, vividly experienced as a subjectively real percept. Others see colored auras around people, with each person having a characteristic color, driven by the synesthete's emotional response to that person. This therefore doesn't represent some kind of psychic ability, though it may well be the origin of such claims. For others, particular flavors induce a tactile feeling of shapes, either felt in the mouth (like little cubes, for example) or in their hand (like a smooth marble pillar)—again, these are very subjectively real tactile sensations.

But there are many other types of synesthesia that have a less florid presentation. Indeed, many synesthetes do not literally experience the concurrent percept as if it were real and projected out in the world; instead, they may experience it "in their mind's eye" or, for many, it may just be a strong and stable association—an extra attribute in their concept of the inducing object.

For example, one of the most common forms of synesthesia is having specific colors for letters of the alphabet (or numbers or time units like days of the week or months of the year). Some synesthetes literally see the color of each letter projected onto the text they are reading (so that everything looks like the Google logo, with each letter having a characteristic color). But, for others, these colors are not seen—they are *known*. In the same way that we know that something with the shape "A" is associated with certain sounds, people with letter-color synesthesia also know that it is red, or blue, or chartreuse, or puce.

Another common type is to associate numbers with very definite positions in space, creating an idiosyncratic "number form." For example, 0 to 10 may run in a straight line left to right in front of the person, then 11 to 20 may go vertically, with a zigzagging line to 30, 40, 50, and so on. Others may have a helical shape or may even have some numbers behind them. Thinking about months of the year in a similar definite spatial arrangement is also a common form. Other properties can also be combined into the concepts of letters, numbers, or calendar units, such as tastes or personalities. One synesthete in a study we conducted described "R" as being "off to the right and tasting of cooked carrots," while August was "like wallpaper and tastes like cream cheese." Another reported that the number "seven is violet with a pleasant, kind personality. Eight is red and has a stronger, less kind personality. Eight is often angry at seven."

Letter-color synesthesia and number forms were described in detail by Francis Galton in his 1883 book *Inquiries into Human Faculty and Its Development*. Given his obsession with statistics, it is not surprising that he characterized these in considerable detail, noting trends in the particular pairings, such as the letters "O" and "I" being almost always either white or black (more on those trends later). Synesthesia became quite a trendy topic in psychology at the turn of the nineteenth century and early twentieth century, but fell out of favor with the field's turn toward behaviorism. This was a movement, led by B. F. Skinner and others, to establish psychology as a respectable and rigorous science. This meant studying only things that were quantifiable in some way, with an emphasis on measurable behavior, and a strong move away from qualitative descriptions of subjective experience. As a result, research into synesthesia declined and there were almost no studies published on it for 60 years or more.

That changed in the 1990s as researchers like Vilayanur Ramachandran, Simon Baron-Cohen, and others resurrected the topic and helped reintroduce it to psychology and neuroscience. Since then we have learned that what was initially thought to be a very rare condition (1 in 20,000 people by one early estimate) is actually far more prevalent— indeed common. As many as 2%–4% of the population may have some form of synesthesia. The facts that synesthesia is effectively benign and not a clinical condition, and also that it is so inherently subjective, have probably contributed to it not being more widely recognized. Most cases of synesthesia are developmental in origin—that is, synesthetes say they have always been that way, rather than having acquired it through injury or drug use. Indeed, many people I have spoken to about it had not realized that the way they perceive the world or think about things like days of the week or numbers or letters was any different from anybody else.

There are many famous examples, especially among artists and musicians, including many whose synesthesia influenced their art in some way. The classical composers Sibelius, Messiaen, and Liszt were all music-color synesthetes, and described using their synesthetic colors in composition. Liszt famously cajoled his musicians, "Oh please, gentlemen, a little bluer, if you please! This tone type requires it!" Or, "That is a deep violet, please, depend on it! Not so rose!"[2] Modern artist Pharrell Williams has said of his synesthesia that "It's the only way that I can identify what something sounds like. I know when something is in key because it either matches the same color or it doesn't. Or it feels different and it doesn't feel right."[3] Other musicians whose work has been influenced by their synesthesia include Duke Ellington, Kanye West, Tori Amos, Billy Joel, and many others.

Many painters and visual artists have been similarly driven by a desire to convey the sense of their subjective synesthetic experiences. Russian painter Wassily Kandinsky saw shapes and colors in response to music (among other forms of synesthesia) and many of his works were aimed at depicting those kinds of experiences. He described, on hearing Wagner's music for the first time: "The violins, the deep tones of the basses, and especially the wind instruments at that time embodied for me all the power of the pre-nocturnal hour. I saw all my colours in my mind; they stood before my eyes. Wild, almost crazy lines were sketched in front of me. I did not dare to express that Wagner had painted 'my hour' musically."[4]

The novelist Vladimir Nabokov describes in his autobiography *Speak, Memory* the colors of his alphabet in exquisite detail: "In the brown group, there are the rich rubbery tone of soft g, paler j, and the drab shoelace of h . . . among the red, b has the tone called burnt sienna by painters, m is a fold of pink flannel, and today I have at last perfectly matched v with 'Rose Quartz' in Maerz and Paul's *Dictionary of Color*." This kind of effort to convey specific synesthetic colors with an aching degree of precision is very typical of synesthetes generally (though they are not all blessed with Nabokov's powers of description). He also relates the moment he and his mother learned of their shared synesthesia,

[2] R. E. Cytowic and D. M. Eagleman, *Wednesday Is Indigo Blue* (Cambridge, MA: MIT Press, 2009), 93.

[3] Pharrell Williams, "On Juxtaposition and Seeing Sounds," *The Record*, National Public Radio, December 31, 2013, https://www.npr.org/templates/transcript/transcript.php?storyId=258406317.

[4] W. Kandinsky, "Reminiscences," in *Kandinsky: Complete Writings on Art*, ed. K. Lindsay and P. Vergo (New York: Da Capo, 1982), 364.

writing, "We discovered that some of her letters had the same tint as mine, and that, besides, she was optically affected by music notes."[5]

This highlights the key fact that synesthesia tends to run in families. Over 40% of synesthetes report a similarly affected first-degree relative. Of course, this leaves 60% of cases as "sporadic" but, as described for neurodevelopmental disorders (see chapter 10), such cases can also have a genetic cause if they are due to new mutations. In families with multiple synesthetes, the pattern of inheritance appears strikingly Mendelian—most consistent with causation by a single dominant mutation, which some family members inherit and others don't. As with some of the other conditions mentioned above, the identities of the particular genes involved are not yet known, though efforts are currently under way to find them. I say "genes," plural, because, while it seems a single mutation is at play in each family, all the evidence to date suggests these are not all in the same gene across different families.

One of the more striking findings from these kinds of genetic studies, including one carried out by my colleagues Kylie Barnett and Fiona Newell and myself, is that different forms of synesthesia are observed in different members of the same family (as Nabokov also noted for his mother's musical form). This means that what is inherited is a predisposition to develop synesthesia in a general sense, but the precise form that emerges in any individual is not so tightly determined. This raises two key questions: What might the functions of those genes be? And what other factors determine the precise outcome in any individual?

The answer to the first question remains speculative, but the genes involved are thought to affect the organization of cortical circuits in some way. Synesthetic experiences are most readily explained if one cortical circuit that processes one type of stimulus (say, sound) in some way *cross-activates* another one, which mediates some additional percept (say, vision). This suggests it could be caused by mutations that affect either the establishment of cortical circuits or the way that they communicate with each other. Whereas the agnosias may be due to a deficit in the integration of extended subregions into dedicated circuits, synesthesia could be due to a failure in the *segregation* of such circuits from each other.

---

[5] Vladimir Nabokov, *Speak, Memory* (London: Victor Gollancz, 1951), 35.

THE EMERGENCE OF THE PHENOTYPE

If a mutation were affecting how connections are formed between cortical areas in the developing brain, or how they are pruned away over time, this might manifest in a probabilistic way, independently, across the brain. Most mutations affecting neurodevelopmental processes tend to have that characteristic—they change the *probability* of one outcome over another, rather than switching from one to the other completely, in every cell or every region. This is seen in mutations affecting cell migration in the cortex, for example, where a tendency to have groups of cells in the wrong place is inherited, but the precise locations where this happens are essentially random. Similarly, some types of epilepsy are highly heritable, but the anatomical locus of seizure activity varies considerably, even between MZ twins. The same logic may apply for synesthesia, where differences in type have also been observed even between MZ twins. There may be a certain probability for excess connections to be formed or retained between any given cortical areas, but the precise outcome in any individual will depend on how that probability plays out across the brain.

Experience could also conceivably play a role in determining the type of synesthesia that someone has, but we simply don't know at the moment whether that's actually true or not. Where there *is* evidence of a role for experience or learning is in biasing or even driving the specific pairings of inducing stimuli and concurrent synesthetic percepts. If you look, for example, across large numbers of synesthetes with colored alphabets, you can see some interesting trends.

The most striking thing is actually the overall arbitrariness of the associations—for most people for most letters, there is no obvious explanation for why it is that particular color for them. However, across many English-speaking synesthetes, we can see that, for example, the letter "Y" is perceived as yellow by about 50% of them, while "R" is red for about 30%—both significantly more than you'd expect by chance. This suggests an influence of semantic associations on the specific pairings that emerge. On the other hand, "Q" is somewhat more commonly purple than expected, "J" is more often orange or brown, and both "O" and "I" are almost always either white or black. It is much harder to see how semantic associations could be driving those effects.

Nevertheless, these trends do highlight an important fact (along with others, such as the numbers 1 to 12 in number forms reported by Galton often being perceived spatially in a clock-face arrangement, or particular words tasting of similar-sounding foods—"Barbara" tasting of rhubarb, for example). Synesthetic pairings emerge over time, as the inducing stimuli themselves (letters, numbers, words) are being learned. We obviously can't think "A" is red until we have a concept of what "A" is. Ordinarily, that concept is built up in regions of the brain—in the anterior inferior temporal (AIT) cortex—that receive inputs from both visual and auditory areas. The shape of the grapheme "A" and the sound of the phoneme |ā| are represented in the patterns of activation of neurons within the visual and auditory areas. As those patterns get repeatedly coactivated—when a child sees the grapheme and hears the phoneme at the same time, over and over again—those two attributes get linked in the AIT region and incorporated into the schema or conceptual representation of the letter "A" (see figure 7.5).

In someone with letter-color synesthesia, we can imagine a scenario where their "color area" of the cortex is getting cross-activated by connections from either the visual form or auditory regions. Whatever pattern represents a particular grapheme or phoneme may then be transferred to the color area to generate some arbitrary color percept— the brain simply interprets that activity as *meaning* red or purple or green. Now, because that color percept will also be reliably coactivated with the shape and sound percepts, it too will be incorporated into the schema of the letter and generate a deeply felt association.

That kind of idea can explain the arbitrariness and idiosyncrasy of most synesthetic associations. But, importantly, it also leaves room for semantic effects because the process takes place over such an extended period of time. If, every time a synesthete sees the letter "Y," that individual is semantically primed to think of the color yellow, then that top-down effect may override the color percepts that the brain is trying to attach to the letter, leading to a bias in associations across many synesthetes. For some synesthetes, particular colors of the alphabet may even be driven by external objects, such as sets of alphabetical refrigerator magnets. A subset of synesthetes, who were born in the United States in the early 1970s, have identically colored alphabets that clearly correspond to one particular set of such magnets that were

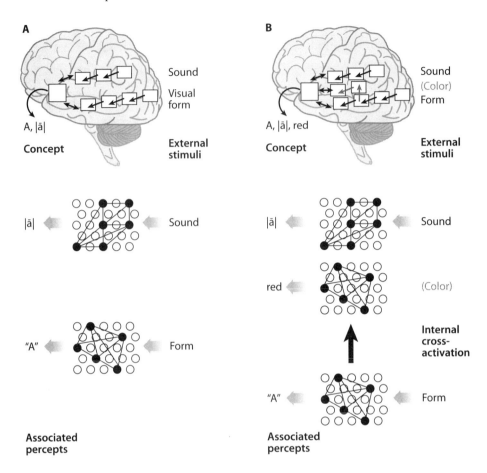

Figure 7.5 Synesthesia. **A**. As we learn letters, we associate the sounds they make with their shapes, and form a higher-order concept of each letter. **B**. In people with letter-color synesthesia, the color area may be internally cross-activated, leading to an arbitrary but consistent color percept, which becomes incorporated into the concept of the letter. (Semantic associations, such as "B is for blue," may sometimes override arbitrary color percepts and be consolidated as synesthetic associations.) (Modified from F. N. Newell and K. J. Mitchell, "Multisensory Integration and Cross-Modal Learning in Synaesthesia: A Unifying Model," *Neuropsychologia* 88 (2016): 140–50.)

a popular children's toy at the time. For what it's worth, I had that toy myself as a child, but couldn't tell you now what color each letter was! Experience can thus clearly bias or even drive the synesthetic pairings that emerge, but there is no reason to think that it drives the condition of synesthesia itself.

WHAT'S HAPPENING IN THE BRAIN?

With all the tools of modern neuroscience at our disposal you might think we would have a good idea what is going on in the brains of synesthetes. In all honesty, though, we just don't. There have been many studies using neuroimaging—including fMRI, EEG, and PET (positron emission tomography) studies—that have tried to find some patterns of neural activity that correlate with synesthetic percepts. And there have been many reported positive findings; for example, of activity in the color area (called V4 or V8) in response to sounds or letters, in sound-color or letter-color synesthetes, respectively (but not in control subjects). These have been taken as an objective indicator of neural activity that underlies the subjective synesthetic experience of color. And that makes sense, in that if you activate that color area with an electrode people do indeed see little blobs or flashes of color "out in the world." So if it was cross-activated internally, you might get a comparable perceptual experience.

However, many other studies have *not* found that signature, despite similar experimental design. Some have seen extra activity in other visual areas, others have not detected any additional activity, and others, including one from my own group, have even seen *decreases* in activity in synesthetes that are hard to interpret as driving an additional percept. The reasons for this lack of consistency remain unclear. A lot of it may have to do with the fact that most of these studies had small samples, which are more likely to generate variable and possibly even spurious results. But it is also likely that synesthetes represent a highly heterogeneous group and that what is true for one small sample may not hold for others.

Many studies have also looked at the structures of synesthetes' brains, compared with controls, to see if they have any consistent differences that correlate with the condition. The hope was that this might distinguish between two hypotheses of what is causing the condition. On the one hand, there might be more connections between some cortical areas than are normally present (a structural difference). On the other, the connections might be present in everyone but working differently in people with synesthesia (a functional difference). Again, there have been many reported findings but nothing that rises to the level of what you would call a "fact." Where differences were observed—for example, in gray matter volume, white matter volume, or measures of structural connectivity in various little regions of the brain—it was typically observed that the synesthetes showed *increases* relative to controls. However, the precise locations of these differences are not very consistent across studies, making it hard to draw definitive conclusions.

Pharmacology hasn't made much inroad into the underlying mechanisms either. There are some drugs that can cause synesthetic-like experiences, but the emphasis is on -*like*. These include well-known hallucinogens like lysergic acid diethylamide (LSD, or "acid"), as well as the chemically active constituents of magic mushrooms (psilocybin) or peyote cactus (mescaline). All of these can induce "trippy" states where, for example, sounds can trigger visual percepts or other sensory phenomena. However, these experiences differ significantly from developmental synesthesia, in that they tend to involve more florid and complex visual forms (objects, people, scenes) and are far less consistent in pairings between inducing stimuli and concurrent percepts.

Despite that difference, the fact that all of these drugs affect the serotonin pathway in the brain has led to proposals that altered activity of these pathways might underlie developmental synesthesia. Additional support for that idea comes from observations that selective serotonin reuptake inhibitors (SSRIs), commonly used as antidepressants, can suppress synesthetic experiences. My colleagues Francesca Farina and Richard Roche and I recently reported a similar case, where synesthetic experiences (in this case colored auras around people and colored music) were completely suppressed for over eight years, while the subject was on SSRIs, yet returned to premedicated levels when the medication was ceased.

However, the same study found that numerous other medications (in that subject and another) also affect the conscious experience of synesthetic percepts. These included diverse drugs with completely different mechanisms of action targeting entirely different neurochemical pathways. The specificity of the link with serotonin is therefore called into question. Like many other conscious experiences, synesthetic experiences may be modulated by many different drugs, without those effects necessarily relating to or informing on the *origins* of the condition.

Though not itself considered a clinical condition, or even a symptom of one, an increased rate of synesthesia is associated with some other conditions, most notably autism. Rates of synesthesia in people diagnosed with autism spectrum disorder have been reported at 17%–18% in two separate studies, much higher than the general population rate of 2%–4%. That relationship may hold in the opposite direction too, at least partly. A recent study by Jamie Ward and colleagues found that synesthetes showed a pattern of sensory hypersensitivity that was very similar to that seen in people with autism, though they did not share the social or communicative symptoms.[6] This relates to other findings, including some from my group, that suggest synesthetes may have more general and more basic differences in sensory processing, beyond those associated with the experience of synesthesia itself.

These differences, in both autism and synesthesia, may affect not just the sensitivity of sensory processing but also the emotional import of various stimuli, the interest they hold and how strongly they are perceived as pleasant or unpleasant—various stimuli may really be *felt* very differently by people with these conditions. In autism, these kinds of basic differences in perception may also contribute to the narrow, focused interests that are one of the cardinal features of the condition.

The combination of narrow interests with additional synesthetic perceptions may also underlie some forms of exceptional or savant abilities that some people with autism display. Many people with autism (estimates range from 10% to 30%) have some kind of exceptional and highly specific talents or "islands of genius." These include, for example, the ability to perform extraordinary mental calculations, exceptional memory, rapid estimation of the numbers of objects, calendar calculations (rapidly deducing that October 11, 2250, will be a Friday, for example), prodigious musical or artistic talent, or other skills.

Some of these seem to be based, at least in part, on additional synesthetic percepts, which alter how various types of objects are mentally coded or manipulated. For example, associating numbers with colors or shapes or positions in space can be a powerful mnemonic aid for remembering long sequences (such as the digits of pi to 20,000 places!). Even more intriguingly, perceiving numbers in spatial arrangements may in some mysterious way allow lightning-fast mental calculations to

---

[6] J. Ward, C. Hoadley, J. E. Hughes, P. Smith, C. Allison, S. Baron-Cohen, and J. Simner, "Atypical Sensory Sensitivity as a Shared Feature between Synaesthesia and Autism," *Sci. Rep.* 7 (2017): 41155.

be performed, by simply "seeing" the answer, as opposed to the laborious sequential processes of arithmetic that most people must carry out.

CONCLUSION

The answer to the question posed at the start of this chapter—do we all see the world the same way?—is very clearly "no." There is in fact a rich and underappreciated diversity of perceptual experience, across all the senses, from a simple level of what kinds of stimuli we can detect, and how they are processed, to a much higher level of how we integrate perceptual attributes of objects into schemas. These differences can be profound, in discrete conditions like face blindness or synesthesia, or vary more continuously across people, like the temporal and spatial resolution of perception. They are largely based on genetic differences in the program that directs the wiring or function of extended and highly complex perceptual circuitry in the brain, as well as random differences in how that program happens to play out in each of us. Ultimately, these differences in perception affect not just what we sense or how we sense it, but also what meaning different stimuli have for us and, at a very fundamental and deeply subjective level, how we all think about various things in our world.

CHAPTER 8

THE CLEVER APE

Some kids are smarter than others. Some are braver, some are kinder, some are more talented, some are more athletic—and some are smarter. They start smarter and they stay smarter. That's a rather stark way to put it, and it may not sound very egalitarian. It may come across as deterministic, even fatalistic, implying that intelligence is an immutable characteristic, that it can't be changed by experiences like education. As we will see below, that is not the case at all. But there is no denying that innate differences in intelligence, or intellectual potential, exist. This is no longer a subject that can be argued about in the abstract, or even one that is situated purely on psychological or sociological ground—we now have many insights into the genetic, developmental, and neural mechanisms underlying such differences.

One reason why discussion of the biological basis of intelligence provokes such strong reactions is that intelligence is humanity's defining characteristic. It is what sets us apart from other animals and what has allowed us to colonize and dominate almost all environments on the planet. Most other animals have tightly defined ecological niches—they live in certain areas, with particular types of terrain and vegetation, they have adapted to a specific climate, they have a limited and highly characteristic set of behaviors.

Of course, the downside of such incredible specialization is that if the environment changes, the animal is undone. All that adaptation that was so beneficial before is now a liability. Most animals have specialized so much—in morphology, physiology, sensory systems, and behavior—that they lack the flexibility to adapt to new environments. As a result, their ecological range is limited—they've painted themselves into a corner. There is a reason that the world is not overrun by pandas or manatees or flying squirrels.

We, on the other hand, took a different evolutionary path. The evolutionary lineage of primates led from creatures like lemurs to ones more like modern monkeys to the emergence of apes, and eventually hominids. At each step along the way, brains got bigger. All kinds of other changes happened, of course, too, in individual branches and in individual species, as they adapted to specific niches, but the overall trend in brain size was quite consistent.

This trend continued across early hominids. We know of dozens of species of early humans that lived tens of thousands of years ago and that have since died out, leaving only *Homo sapiens* as the single representative species of this genus on the planet. Over this time there was a general trend for increased brain size and the archaeological record indicates a concomitant capacity for more complex and more flexible behavior, with the emergence of tool making, cooking, cooperative hunting, trade, and even music and symbolic art. Eventually, this process of increasing intelligence along our lineage reached its pinnacle with the emergence of modern humans.

THE ESSENCE OF INTELLIGENCE

At its core, intelligence is the ability to think in more and more abstract ways—to see a specific instance of something and draw larger lessons from it, which can then be applied to other situations, by analogy. We can go from learning that "A causes B" to extrapolating that "things *like* A can cause things *like* B." That power of analogy is at the very heart of our intelligence—it is, in fact, explicitly included in questions on IQ tests, like: "Acorn is to tree as puppy is to _____." The analogy in that example is based on a quite concrete relationship, but, with increasing brainpower, analogies can be made across higher-order properties of *categories* of things or events or situations.

Let me make an analogy. The hierarchical organization of our visual system allows us to extract features of the visual scene of higher and higher order. Each area integrates information from the lower areas and extracts a more complex model of the world—first just dots and flashes, then lines and edges, then shapes and objects, then types of objects—tools, animals, faces—until we get to a stage where we can categorize objects as the same thing—say, a chair—despite seeing it from different angles, and we can recognize multiple different things as being members of the same category, based on their higher-order properties (like having multiple legs and a flat bit to sit on, for example). Our cognitive systems do the same thing. As the cerebral cortex got larger, it led to the emergence of new areas, so that the hierarchy had more levels, each one able to integrate more sophisticated information from lower levels and discern more and more abstract properties.

When we talk about intelligent behavior we mean the deployment of such abilities to recognize the relevant dynamics of novel situations, to anticipate events, to imagine the consequences or outcomes of a range of possible actions. Intelligent beings are not just driven by hardwired instincts or even by learned responses to specific stimuli—they can use the abstract principles gleaned from prior experience to adapt to *new* situations and environments.

At some point in evolution, the increasing ability to think in abstract terms—to have *ideas*—led to, and was reinforced by, the emergence of language. How this happened is a mystery, of course, tied up with the emergence of consciousness itself, which is definitely a topic for another day. But the consequences were profound. Now the advantages of each individual's big brain were massively amplified by the ability to communicate ideas with each other. Now if I learned something useful, I could tell you; if I had a good idea, I could pass it on so everyone in the group benefited. Then children didn't have to relearn everything anew from their own experiences—instead, they could build on the previous hardwon knowledge of their parents, and of others in the group.

Culture was born. And cultural evolution started to interact and collaborate with biological evolution. Where, previously, being more intelligent gave some advantage, now it gave a huge advantage. And the more intelligent we got, the better it became to be *even more* intelligent. This snowball effect meant that we started to be able to transcend the normal rules of natural selection. We made our own niche—the *cognitive niche*. Instead of being selected by our environments at the glacially slow pace of evolution, we had the flexibility to adapt to them on the fly, and eventually to flip the process entirely—now we were in the driver's seat, adapting our environments to our own ends. In the process we changed the selective pressures acting on new mutations, greatly favoring any that further increased intelligence. The only thing that put the brakes on this process of positive feedback is thought to have been a size constraint—our heads became too big for the birth canal. Or perhaps the metabolic costs of our big brains, which use about 20% of our energy, just became too high. However it happened, we ended up with intellects leagues beyond our nearest relatives.

Because of its central role in our evolution, when it comes to variation in intelligence across people today, this seems, more than other traits, to carry a kind of value judgment with it. Unlike many personality traits, where variation is seen as fairly neutral—where it's not obviously, or at least not consistently, better to be, say, more extraverted, or less neurotic—variation in intelligence is not neutral. All other things being equal, higher intelligence *is better than* lower intelligence.

We'll see how this idea influenced the dark policies of eugenics that were widespread across many countries in the twentieth century (and that are, in some places, experiencing a surprising resurgence, though perhaps in a more benign form). Supporters of eugenic policies made the unwarranted extrapolation that a more intelligent *person* is better than a less intelligent *person*. The idea of judging the "quality" or "worth" of a person at all is repugnant (to me at least, though apparently not to everyone), but if one were to engage in such a practice, intelligence is just one of many personality and character traits that we might throw into the mix (honesty, integrity, kindness, courage, and selflessness all spring to mind as equally valuable elements of our humanity). In any case, given the history and attitude of eugenics, it is not surprising that there was and continues to be a strong backlash against the very idea that intelligence is in any way innate.

In what follows, I will try to separate the science from these kinds of extrapolations, though we will return in chapter 11 to the societal implications of the scientific findings and especially to the subject of eugenics. For now, what follows from the discussion of the evolution of the intelligence of humans as a species should be obvious: that difference, between us and other animals, is genetic. Cultural evolution played a central enabling role, but, ultimately, we each have human intellectual capacities because the program for a complex human brain is written into our DNA. It should not be a surprise, then, that variation in that genetic program could exist between people and could contribute to variation in their intelligence. Indeed, it would be a surprise if it didn't.

MEASURING INTELLIGENCE

To determine whether intelligence is affected by genetic variation, we first need a way to measure intelligence. This is what IQ (intelligence quotient) tests or other psychometric tests of cognitive abilities were developed for. The first such tests were developed in France in 1904 by Alfred Binet, under instruction from the French Ministry of Education. They wanted a way to determine which students were progressing well in school and which ones were falling behind. Binet developed a series of questions and puzzles that children could perform and calibrated these by the ages at which children typically became able to do them. Using that method he could compare children and see if they were performing at a level appropriate for their age. The tests were specifically aimed at identifying students who needed additional help in school.

These tests were later revised by American psychologist Lewis Terman, working at Stanford University, to produce the Stanford–Binet test, still in use today. The philosophy behind this test and, especially, its widespread application in American society in the twentieth century was rather different from Binet's original intention. Rather than being used as a marker of level of achievement to spot children in need of remedial education, it was instead used to pigeonhole people, as an indicator of a fixed trait of intellectual potential. We will see below that people do indeed differ in intellectual potential, but absolute levels of intelligence in individuals are by no means fixed.

The Stanford–Binet test measures five factors: knowledge, quantitative reasoning, visual-spatial processing, working memory, and fluid reasoning. It includes questions that require some knowledge; for example, tests of vocabulary and verbal fluency—like how many words can you think of beginning with the letter "P" in a minute? But it also tests abilities that are less reliant on concrete knowledge, such as ability to mentally rotate shapes, to hold spans of digits in your mind, or to detect some patterns or trends (even in nonsensical shapes or symbols) and use them to predict the next element of a series (see figure 8.1 for some sample questions). Some of the tests measure speed of responses, or reaction time, as well as accuracy. Full IQ tests are administered and scored by trained professionals and show high test-retest reliability, of around 0.90. There are, in addition, a number of other tests of general

1. Which of the cubes is the same as the unfolded cube below?

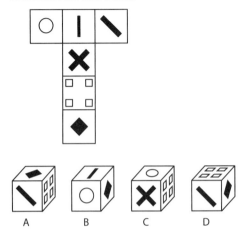

4. Which shape completes the sequence in the bottom row?

2. 19205111 is to steak as 381918 is to:

A. peace B. chair C. prawn D. kite

3. All is to Many as None is to…

A. Never B. Some C. Always D. Few

5. Complete the sequence:
4, 5, 8, 17, 44…

A. 56 B. 68 C. 81 D. 125

Figure 8.1 Sample IQ test questions.

or specialized cognitive abilities that are often used in genetic research, which are much quicker to administer but which show somewhat lower test-retest reliability.

The important thing about all the different factors on the test is that the results tend to be correlated across them. Not perfectly, as each of us has a different profile of strengths and weaknesses across these different factors, but the overall relationship between them is strongly positive. That is, people who have faster reaction times tend to also have greater vocabulary and a better ability to rotate shapes in their mind or perform mental operations on long numbers. That might be surprising, as these different tasks don't seem to be obviously related to each other. And each of them does indeed involve some task-specific faculties and knowledge. But there is also clearly some underlying more general factor that contributes to all of them, which explains the correlation between them. This factor is known as *g*, for general intelligence.

The *g* factor can be calculated statistically and usually accounts for 40%–50% of the variation in performance across each of the different elements of IQ tests. IQ scores thus reflect this *g* factor as well as variation in specific factors. By convention, IQ scores are normalized to a mean of 100, calibrated from large test samples of the population. Most people will score somewhere near to the mean, with fewer and fewer people as you go out to the extremes at either the low or high end of the distribution. We'll come back to the nature of this distribution below.

But first, we should ask: Do IQ tests actually tell us anything useful? Do they only measure how good people are at taking IQ tests, or tests in general? Are they so culturally biased or confounded by environmental variables that no conclusions about biological differences can possibly be drawn from them? Well, it's certainly true that they tap into not just innate differences but also the effects of experience, in particular of education. However, while education does increase performance on these tests, it is still true that some children at the same educational level will perform better than others (that is exactly what the tests were designed to capture). And those *relative* differences in cognitive ability remain quite stable over the lifetime.

Moreover, IQ scores in children are predictive of many real-world life outcomes. These include years of education, academic success (exam performance), income, training success, job performance, general physical and mental health, and longevity. Indeed, in a study of over a million Swedish men, followed over 20 years, the men in the lowest ninth of the IQ distribution had a threefold higher rate of death compared with those in the highest ninth, and this difference extended smoothly across the whole distribution. So, whatever differences IQ tests are tapping into, they are clearly important for how we get on in life.

IQ ACROSS THE POPULATION

The distribution of IQ scores across the population looks pretty "normal," in a statistical sense. That means you have lots of people near the middle, and fewer and fewer as you go out to the extremes—a typical bell-shaped curve. But if you look more closely you can see that the distribution is not perfectly bell-shaped—there is a little bump at the lowest end of the distribution, a slight excess of people with very low IQ

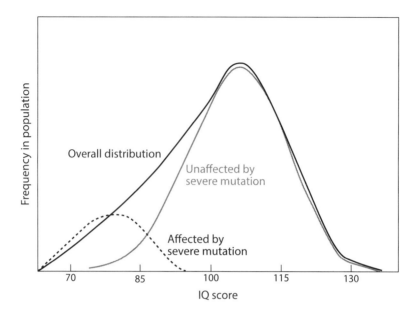

Figure 8.2 IQ distribution. The distribution of IQ can be modeled as a large statistically normal distribution, with a smaller, distinct subdistribution at the low end. This represents people with severe mutations that substantially reduce IQ, sometimes causing clinical intellectual disability. (Summarizing data from Scottish Mental Surveys as presented in W. Johnson, A. Carothers, and I. J. Deary, "Sex Differences in Variability in General Intelligence: A New Look at the Old Question," *Perspect. Psychol. Sci.* 3, no. 6 (2008): 518–31.)

(see figure 8.2). This bump represents people with intellectual disability. The fact that it deviates from the otherwise smoothly normal distribution tells us something very important—the reason they have low IQ is not because they happen to fall at that end of a single continuum—they are exceptional cases with a different underlying explanation.

The vast majority of cases of intellectual disability are genetic in origin. There are well over 500 specific genetic conditions that are known to cause substantial cognitive impairment, and more are being identified all the time. These include chromosomal disorders like Down syndrome, deletions or duplications of segments of chromosomes like Williams syndrome or Angelman syndrome, and conditions where only a single gene is affected like fragile X syndrome or Rett syndrome.

By itself, the existence of these conditions clearly shows that human intelligence is "genetic," in the sense that it relies on a complex program encoded in our genomes and can be seriously affected by mutations that compromise that program. The fact that there are so many ways to disrupt that program shows just how complex it is—severe mutation of any one of hundreds of different genes is sufficient to drastically derail brain development or function.

In some cases, these mutations are inherited from one or both parents, but they often arise as new mutations that occurred during the generation of sperm or eggs. One reason for this is that people with intellectual disability tend not to have children. So, if a single copy of a mutation is sufficient to cause the condition, it means the parents are unlikely to have carried it—otherwise they most likely would not have become parents. Where a mutation is inherited, it is more commonly a recessive or X-chromosome-linked mode of inheritance. For recessive conditions, each parent could carry one copy of the mutant gene, but still have another copy functional—only in the offspring, where two mutant copies can come together, do you get any impairment. Mutations on the X chromosome can often be carried by females without major consequence, because they have a backup X chromosome, but if they pass on the mutant copy to their sons, who only have one X, along with a Y chromosome, then those sons can be severely affected.

When we look at the relatives of people with intellectual disability we see something very interesting. For the most part, the IQ of their relatives does not differ from the mean of the population. That is because the condition is typically caused by a single, discrete mutation (in one or two copies). If this arose de novo, then the relatives of affected individuals should not carry it and should therefore be completely unaffected. Parents or siblings of people with Down syndrome, for example, will usually have IQs in the normal range. For the same reason, the relatives of people with dwarfism tend to have heights in the normal range.

However, when we look instead at the relatives of people who have relatively low IQ, but who are not in the extremely low range where they would get diagnosed as clinically intellectually disabled, we see something different. The relatives of such people *do* have IQs that are lower than average. This suggests that the genetics of intelligence across the typical range differs from the genetics of intellectual disability. Rather than being determined by single mutations, it is much more likely to involve effects of multiple genetic variants at once. Relatives of people with low IQ will share some of these and therefore also have lower than average IQ. Again, for the same reason, the relatives of people who are just short, but who do not have dwarfism, do tend to also be shorter than average.

Now, you might at this stage be thinking: "Hang on there. The fact that the relatives of people with low IQ also have low IQ does not mean that IQ is genetic." And you're dead right—after all, you could replace "IQ" with "wealth" in that sentence and still have a strong correlation between relatives. It could be completely determined by environmental factors that are also shared between family members. Familial correlations by themselves suggest a genetic mechanism but do not prove it. To do that, we have to turn again to analyses designed to separate genetic from family environment effects—twin and adoption studies.

TWIN AND ADOPTION STUDIES OF INTELLIGENCE

If intelligence were completely genetic, then adoptive siblings should be no more similar than any two random people. On the other hand, if it were completely determined by family upbringing, and correlated social factors such as education, then adoptive siblings should be as similar to each other as biological siblings. In fact, we see something in between, at least at first. Many studies have consistently found that adoptive siblings are more similar than random people (a correlation in IQ scores of ~0.25) but less similar than biological siblings (who typically have a correlation of ~0.60). This suggests that both genetic relatedness and a shared family environment can affect intelligence.

However, this is only found when IQ is measured in children. At this young age, when the children are still in the bosom of the family environment, there is indeed an effect—differences in family environments can clearly contribute to differences in performance on IQ tests in children and young adolescents. But when adoptive siblings are tested later in life—in some longitudinal studies, the very same siblings—that correlation vanishes. They are no longer any more similar to each other than a pair of strangers. The similarity between siblings also tends to decrease, slightly, presumably as that effect of a shared family environment also dissipates for them, but this leaves a very substantial correlation (of 0.40–0.50) due to genetic relatedness. Thus, whatever the differences in nurture that influence cognitive performance in the short term, they do not seem to have long-lasting effects.

These findings are borne out by the results of twin studies. These have consistently found that MZ twins are much more similar to each other (correlation coefficient, $r = 0.75$–0.85) than DZ twins ($r = 0.40$–0.50). This is all the more remarkable considering the test-retest reliability of IQ tests is only about 0.90. So, the scores of MZ twins are often nearly as similar to each other as scores for an individual across two testing sessions. These are the numbers when people are tested as adults—when tested as kids, the difference between DZ and MZ twins is less stark. Again, this suggests the temporary effect of a shared family environment, which fades over time. A smaller number of studies have combined these two study designs—looking at twins who either were or were not adopted away from each other. The finding is stark—MZ twins who were adopted away and raised in different families end up nearly as similar to each other in IQ ($r = 0.78$) as MZ twins who were raised together ($r = 0.85$ in these studies).

If we estimate the heritability of IQ from these sorts of data (i.e., the percentage of the variance in the trait that is attributable to genetic differences), we get different answers at different ages. In infants, most of the similarity is due to the shared family environment and heritability is very low. But over time, the familial effect gradually gets smaller and smaller and the genetic effect larger, such that in adults the heritability is 75%–80% and the variance attributed to the family environment is zero.

IMPORTANT ENVIRONMENTAL EFFECTS

On the face of it, these results may seem to imply that intelligence is an immutable, innate trait, that it is not affected by our experiences, and that differences in environment do not have any lasting effect on it. In fact, none of those conclusions is warranted and none of them is true. That is because our ability to draw conclusions about environmental effects from twin and adoption studies is limited to the environmental differences *actually sampled* in those studies. For example, there is some evidence that the heritability of IQ is lower in samples with lower socio-economic status, suggestive of greater environmental variance in low-socioeconomic-status communities.

Because of the way they are carried out, these kinds of studies tend to sample quite a limited range of environmental differences. Subjects are typically recruited from within the same population—often from a limited region—and at around the same time. The design of twin studies tests for effects of differences in family environments but does not test the possible effects of broader environmental factors, because these tend not to differ between families in the study. They tell us only what *does* influence variance in the trait, in the given population studied at that time, not what *could* influence the trait. In particular, they typically do not assess the effects of cultural or societal differences between regions, or countries, or over time.

Differences in IQ scores over time are particularly important. IQ tests have been around for over a century and we have data on performance for different populations over that time. At any particular time, the average performance across the tested population is normalized to 100, by convention. But when you actually look at the *absolute*, not the normalized, scores, across the decades, you see something really striking—these have been consistently increasing over time. It's not that the tests have changed—people are getting better at them, on average. This so-called "Flynn effect," named after its discoverer James Flynn, is an incredibly consistent observation—seen across essentially all countries where data are available.

Exactly what's causing it is a matter of debate, but it probably involves multiple factors that have changed over time. These include better nutrition and generally improved maternal and childhood health, which presumably favor optimal brain growth and development. They also include better and longer education (more on that below). And, more generally, the effect may reflect an increasing trend toward abstract habits of thought in modernizing societies. As science and technology have progressed, and as industries, jobs, and a huge range of other societal factors have changed, the ways we spend our time and the things we spend time thinking about have also changed.

The one thing that is clearly *not* an explanation of the Flynn effect is changing genetics. There have simply not been enough generations in the time span over which these increases in IQ have been observed. The differences are definitively environmental in nature. This re-emphasizes a hugely important point—a trait can show very high heritability in any given population at any given time and still be affected by environmental differences *between* populations, as observed for body mass index, for example.

This becomes important when interpreting differences in average IQ between different populations or subpopulations. There are, for example, data showing that African Americans or Hispanic Americans have lower average IQ scores than European Americans. Because intelligence is highly heritable, this has been interpreted by some as evidence of a genetic difference in intelligence between ethnic groups. In fact, that conclusion is not warranted. We know that differences in nutrition, general health, and education can all strongly influence IQ scores. The fact that such socioeconomic differences exist across ethnic categories in the United States thus suggests a plausible explanation—at least as likely as, and indeed, more parsimonious than, invoking underlying genetic differences.

An example from Ireland illustrates this fact. In the 1970s the average IQ score in Ireland was around 85—a massive difference from the average of 100 seen in the United Kingdom at the same time. This was taken as evidence that the Irish were constitutionally stupid—not just ignorant and poorly educated, but irredeemably simple. But things were changing in Ireland at the time. It went from an overwhelmingly rural, agricultural society to one with increasing urbanization, industrialization, and prosperity, with concomitant increases in nutrition, health, and average length of education. By the mid-1990s, IQ scores were averaging around 95 and they now average 100, exactly on par with our neighbors in the United Kingdom. Nothing changed genetically over that time— better circumstances just allowed that latent potential to flourish.

And, indeed, education does exactly the same thing in all of us. As we get educated, we don't just learn new things, *we get smarter*. We assimilate ideas, not just facts, and we begin to deploy those concepts in new situations—the very definition of intelligent behavior. And we get progressively better at learning new things, at making new associations and grasping new and more complex ideas. That is why IQ tests are calibrated for people of different ages—in fact, it was specifically to test how well children are progressing that IQ tests were first developed. While differences in intellectual potential may be innate, actual intelligence increases over time, with maturation and education.

However, this does not mean that everyone will reach the same point. While absolute intelligence increases over childhood in each of us, our *relative ranking* remains remarkably stable. Children who had higher IQ scores at age 5 will tend to still have higher IQ scores at age 20 or age 30, even though everyone's cognitive abilities will have increased over that time. In fact, longitudinal studies have shown that IQ scores taken at age 11 are very good predictors of relative rank in the same cohort at age 87. A rising tide lifts all boats, but some may still sit higher in the water than others.

Moreover, while increasing access to education benefits everyone, it may not do so evenly. Those with higher initial IQ may benefit more from education—they may learn more readily and be able to apply that knowledge more productively. They may find education more interesting and rewarding and may therefore apply themselves more. They are likely to be more encouraged by parents and teachers and thus choose to stay in education longer. This means that while more educational opportunities will increase everyone's intelligence, those who start at the higher end may benefit the most. Rather than simply shifting the whole distribution upward, greater education may actually exaggerate initial differences.

The fact that MZ twins tend to get *more similar* to each other in IQ with age illustrates the experience-dependent amplification of this trait. Perhaps more than any other trait, intelligence has a progressive directionality to it—the more you learn and understand, the easier it is to learn and understand even more. The fact that heritability of intelligence increases with age thus reflects not just a diminishing of the temporary effect of the family environment, but an amplification of innate genetic differences in intellectual potential.

All of the preceding presents the evidence for something that has been known for a long time—that intelligence is highly heritable. This shouldn't be a surprise—it certainly jibes with common experience that some kids are innately more intelligent than others and that this trait runs in families. But we can now go far beyond that simple conclusion. In recent years, we have been able to ask what kinds of genetic variants underlie this trait and have even begun to identify individual genes involved. These discoveries are beginning to illuminate the biological basis of intelligence and the kinds of brain parameters it reflects.

THE GENETIC ARCHITECTURE OF INTELLIGENCE

The fact that intelligence is a continuously distributed trait, with values smoothly spread across a range, and the observation that it shows blended rather than discrete inheritance indicate the involvement of multiple genetic variants in any individual. One way you can get that kind of distribution is if there exists a limited set of common genetic variants that segregate in the population that either increase or decrease intelligence. We might call them "plus" and "minus" variants. If you happen to inherit more plus than minus variants, you'll end up more intelligent than average, and, conversely, more minus than plus variants will put you on the other side of the distribution. Parents who carry many plus variants will have children who also have more plus variants. Siblings of people with low IQ will share an excess of minus variants and likely have lower than average IQ themselves.

That is the standard model of what is known as quantitative genetics— the genetics of traits that can be measured across a range, like height, as opposed to more discrete traits, like eye color. The idea of a standing pool of genetic variants affecting a trait is very useful in animal or plant breeding. If you have a large herd of cattle that differs in, say, milk yield, you can selectively breed together the ones at the high end of the distribution and enrich for the plus variants while gradually eliminating the minus ones, increasing milk yield generation by generation. (If you're wondering how you select bulls on milk yield, it's done indirectly by looking at their sisters or daughters.)

There are, however, a number of reasons to think that this standard model of quantitative genetics does not apply to intelligence, or at least that it is both simplistic and potentially misleading. First, it assumes a static pool of genetic variation, but, as we have seen, the spectrum of genetic variation in human populations is actually highly dynamic, with new mutations entering the population all the time and others being eliminated. We are not a herd of cattle—humanity's recent population explosion has introduced vast numbers of new genetic variants to the population. Second, it assumes a combined effect of very many variants per person, with each one having only a small effect by itself. A combined effect is certainly likely, but we have seen with the examples of intellectual disability syndromes that individual mutations can also have very large effects on intelligence. There is no reason to think that the effects of such mutations should be restricted only to the extreme end of the distribution. Third, the quantitative genetics model explicitly ignores the important effects of natural selection—both its current role and its effects in the past in shaping the genetic architecture of the trait.

By considering a more dynamic view, that incorporates both previous and ongoing mutation and selection, we can derive a very different set of expectations for the genetics of intelligence. The first of these expectations is that "plus" variants—ones that increase intelligence—should be vanishingly rare. Evolution has crafted a finely honed machine—the human brain—over hundreds of thousands of years. We have even become complicit in this selection program, with the inventions of language and culture making it ever more beneficial to be even a small bit smarter. As a result, the space of possible mutations that would increase intelligence has likely already been exhaustively explored by natural selection. New mutations that did so would have been strongly selected for and most likely rapidly become fixed in the population, replacing the previous version of whatever gene they affected.

That's not to say that it's impossible for a new mutation to increase intelligence even further—just that it's highly improbable. Indeed, there are some experimentally induced mutations that are known to increase learning and memory capabilities in animals like flies and mice. But generally, if a new mutation has any effect on intelligence at all, it is far more likely to reduce it than to increase it. It's just much easier to mess up a complex system than to improve it. Imagine taking a wrench to random parts of a Formula One car engine—there's probably not much, maybe not anything, you could do to it that would improve its performance.

From that perspective, we can say that the genetic architecture of intelligence is likely dominated by "minus" variants. These are not genes "for intelligence"—quite the opposite, in fact. Perhaps what we're really talking about is the genetics of stupidity. Under this model, the distribution of intelligence would reflect how many minus variants we each carry—how far we each are from what we might call the Platonic ideal, the theoretical maximum intelligence of a fully wild-type human. Of course, such humans don't exist and never have—all of us carry hundreds of rare genetic variants that impair the production or function of some protein, as did all of our ancestors, along with many thousands of other genetic variants of smaller effect.

Having created such a precision instrument, natural selection now has the massive job of protecting it. Positive selection led to the fixation of genetic variants that increased human intelligence. Negative selection keeps them that way, or tries to at least. This is a constant battle, as new mutations arise with each new sperm or egg. Mutations that drastically reduce intelligence are rapidly selected against, because people with a clinical level of intellectual disability tend to have far fewer children, if any. But mutations that have a more subtle effect are harder for natural selection to deal with. There are, in the first instance, simply too many genes involved in building a complex human brain for natural selection to keep an eye on all of them at once. Weak mutations can sneak through and may drift to relatively high frequencies if their effects are small enough. Even ones with moderate effects may evade natural selection for a while and become new variants in the population.

And there may be another reason why such minus variants would persist, at least in modern times. There is every reason to think that in humanity's distant and not-so-distant past, greater intelligence was selected for and lower intelligence was selected against, both in terms of survival and reproductive success—which is measured not just by number of offspring, but also number of offspring who themselves survive to breeding age and have children. That may be changing a bit in modern times. Intelligence is still negatively correlated with mortality, from all kinds of causes, with lower intelligence associated with greater risk of death from cardiovascular disease, respiratory disease, many cancers, infectious diseases, and other natural causes of death, as well as from accidents, homicide, and suicide.

However, the correlation with reproductive success no longer holds. Higher intelligence correlates with a later parental age at birth of the first child and a smaller number of children overall. As rates of infant mortality have plummeted over the past couple of hundred years, and evened out over the socioeconomic spectrum, the number of children

born now maps much more directly to the number of children surviv-
ing to adulthood. The fact that people with lower IQ tend to die earlier
is thus offset by the fact that they have more children and have them
younger. This effect can thus contribute to the persistence of variants
with weak negative effects on intelligence in the population.

So, what all of this means is that there are lots of genetic variants in
the population that can lower intelligence and each of us carries some
burden of them. We all have some set of rare variants with possibly large
effects, as well as a burden of more common variants with small individual effects, but a considerable collective influence.

FINDING THE GENES

Scientists have been searching for decades for "the genes for intelligence." Only in recent years, with the development of new technologies,
has this search started to pay off. As discussed above, however, it is clear
that we should more accurately think of the discoveries as "genetic variants that affect intelligence," mostly negatively.

On the rare variant front, genomic technologies have allowed researchers to identify recurrent copy number variants (CNVs)—deletions or duplications of segments of chromosomes that change the number of copies
of the genes within those segments from the normal two to one, for deletions, or three, for duplications. These recur over and over again at specific
sites in the genome due to the presence of repeated sequences of DNA,
which confuse the machinery that recombines chromosomes during cell
division, especially during the generation of eggs and sperm. Many of
these CNVs are associated with high risk of neurological or psychiatric
conditions, including intellectual disability, autism, epilepsy, schizophrenia, and other disorders. We will see more of these in chapter 10.

But not everyone who carries such a CNV develops such symptoms—
many are clinically unaffected. However, they may not be totally unaffected. The presence of such CNVs is associated with decreased cognitive
ability and lower performance on IQ tests—not to the level that would
lead to a diagnosis of intellectual disability, but a reduction of anywhere
from 5 to 20 IQ points. While these particular deletions or duplications
are only carried by ~1% of the population, they illustrate a more general

point—that rare genetic variants can contribute to variance in intelligence not just at the extreme low end, but across the distribution.

Because those CNVs recur at specific sites, they are easy to detect and their effects can be readily measured across multiple subjects. It is far more difficult to assess the effects of single base changes in the DNA sequence (known as single-nucleotide variants, or SNVs) because these occur at random across the whole genome, making each individual one exceedingly rare. But it is possible to assess their effects as a group. SNVs can be ranked based on how severe their effects are likely to be on production or function of a protein. Ones with a large effect tend to be kept rare by natural selection. But we all carry some of them and some of us carry more than others. It is therefore possible to look generally at the burden of this class of rare genetic variants and ask whether it is associated with any differences in intelligence.

The answer appears to be yes, though these kinds of analyses have only begun to be possible in the past couple of years. One study found that the number of "ultrarare" genetic variants (ones seen in only 1 person in a sample of over 70,000) was negatively correlated across individuals with educational attainment in a general population sample of over 14,000 people. Educational attainment is used as a proxy for intelligence because it is very easily measured in huge samples (you just ask people how far they progressed), is around 40% heritable, and is correlated with IQ (and also weakly with some personality characteristics including conscientiousness and openness to experience). This finding suggests that ultrarare variants can indeed have an important impact on intelligence, across the general range. As more and more people have their genomes sequenced, larger studies of this type will become possible, enabling greater definition of the effects of rare variants.

Common genetic variants play an important role too. Massive genome-wide association studies, now with hundreds of thousands of subjects, have begun to identify individual variants that are significantly associated with either performance on cognitive tasks directly or with some proxies for intelligence, including educational attainment or head circumference. These studies examine the frequencies of the different versions at sites in the genome where there is a common variant (i.e., the DNA letter at that position varies across people, such that, say, 30% of people might have an A there and 70% might have a G). In this case, they looked for sites where the frequency of one of the variants was associated with degree of educational attainment (so that, say, the A version in the example above went from 27% frequency among people at the lowest end to 33% at the highest end of the distribution of education).

This study found 74 common variants showing that kind of pattern, which replicated in another sample. Each of these has a tiny statistical effect by itself—the differences in frequency tend to be quite small and the trends are only visible above the background of random variation when sample sizes get really huge. By looking at all of these variants in combination we can get an indication of their collective effect. For now, that remains pretty weak—collectively the information from these variants predicts only ~3% of educational attainment variance. But it's a start, and there is reason to think that the combined effect of all common variants (including all the ones not yet individually identified) is much larger—explaining maybe 30% of variance in educational attainment.

It is still very early days in this field. We have only begun to find these kinds of genetic variants, rare or common, that influence intelligence across the typical range. Progress has been extremely rapid over the past couple of years and, by the time you read this, many more variants will likely have been identified. But even with the incomplete picture we have now, we can draw a number of important conclusions about the types of genes involved.

GENES FOR BUILDING A BRAIN

The most striking thing about the genes implicated, both in rare and common variants, is that they are strongly expressed in the brain, particularly in the fetal brain. And many of them—far more than expected by chance—encode proteins with roles in neural development, including neuronal proliferation, cell migration, guidance of growing axons, synapse specification, and synaptic plasticity. It's important to emphasize that this didn't have to be the case. First, if the genetic findings were spurious—just noise mistaken for signal—then there should be no enrichment for genes with brain expression or neural functions. So, these results strongly indicate that the genetic findings are real.

Second, the fact that the genes are enriched for roles in neural *development*, as opposed to genes involved in the function of the mature brain, is particularly remarkable. Again, this didn't have to be the case. One could imagine that differences in intelligence might result from differences in things like the efficiency of brain metabolism, or the activity of particular neurotransmitter pathways, or the balance of various ion channels. Variation in all kinds of biochemical parameters in the mature brain might conceivably underlie differences in either general function or the function of some hypothetical specific circuits mediating higher cognition.

But that appears not to be the case. The variation that we see is instead in genes controlling how the brain was put together. That's a hugely important finding as it suggests that intelligence is not linked to specific brain circuits or specific neurotransmitter pathways. Instead it may reflect general parameters of the robustness and computational efficiency of the brain's networks. This fits completely with results from neuroimaging studies that have looked for correlates of intelligence in the structure or function of the brain.

CORRELATES OF INTELLIGENCE IN THE BRAIN

What neuroimaging studies have found is that intelligence correlates far better with *global* measures of brain structure or function than with parameters relating to any specific circuits or regions. The most obvious brain parameter that correlates with intelligence is size—overall brain volume shows a correlation with IQ of about 0.40. When people have looked more closely to see if the volume or thickness of specific regions of the brain, especially of the cerebral cortex, correlates more strongly with IQ, they have not really been able to narrow this down dramatically. Regions in the frontal and parietal lobe show somewhat stronger correlations with IQ, but these are by no means exclusive. In general, the effects are distributed across most of the cortex.

This is true too when looking at the structure of brain connections. IQ is correlated with the integrity of white matter tracts across the whole cortex. At a global level, it is especially correlated with measures of the "efficiency" of the structural network of fibers—how well connected the entire network is, as opposed to the connectivity of specific areas.

Functional parameters show the same pattern. Studies measuring neural activity with electroencephalography (which involves detecting electrical currents in the brain with electrodes on the scalp) or fMRI have consistently found that IQ correlates better with global measures of network efficiency than with the function of any specific area or the strength of any specific connection. One fairly consistent finding is that during tasks that are reasonably difficult, higher intelligence correlates with *lower* levels of brain activation of various cortical areas. That may sound a bit surprising, but the interpretation is that the brains of more intelligent people have to work less hard to accomplish the same task.

Another consistent finding, from neurology, is that the starting level of a person's intelligence is a decent predictor of recovery after a brain injury, such as a stroke, or of the rate of decline in conditions like Alzheimer's or Parkinson's disease (with higher initial IQ being a protective factor). This has been referred to as "cognitive reserve"—a measure of the brain's resilience to insult. The fact that this correlates with IQ reinforces the notion of intelligence as a general indicator of the robustness of brain systems. The genetic and neurobiological findings thus paint a consistent picture—intelligence reflects how well the brain is put together, how robust the genetic program of neural development was, and how efficient the resultant neural networks are.

IS INTELLIGENCE A GENERAL FITNESS INDICATOR?

One other striking characteristic of the genes so far implicated in variation in intelligence is that they show signatures of being strongly under the influence of negative selection. Natural selection keeps a closer eye on some genes than others, such that the sequences of some genes are especially highly constrained. This is based at the biochemical level on how important it is for the encoded protein to have a particular sequence, but, more importantly, it is based at the organismal level on how important it is to have that protein expressed normally or functioning normally. For the genes implicated in intelligence, it is clear that variants that impair function reduce evolutionary fitness and are gradually removed from the population, or at least kept at a low frequency (the strength of selection being proportional to the size of the effect of the variant).

However, there is more than one way to think about this. Until now, I've been talking about natural selection acting on intelligence itself—that is, being more or less intelligent is the thing that affects survival or number of offspring (and their survival). But there is another scenario where intelligence is not the crucial evolutionary factor itself—it is merely an *indicator* of more general effects on fitness. In this model, the overall load of deleterious rare variants we all carry, interacting with a wider background of less severe genetic variants, impairs developmental robustness generally (see figure 8.3). This could arise—indeed we know it does arise—as multiple deleterious variants accumulate. When one component of the developmental system is disrupted, it affects not just the processes it is directly involved in, but also the ability of the system to compensate for other changes or for the ubiquitous molecular noise

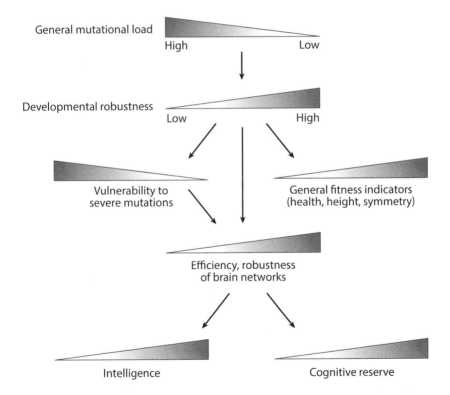

Figure 8.3 The genetics of intelligence. A model incorporating general effects of mutational load on developmental robustness, and its possible impact on intelligence.

in the system. As a result, the robustness of the system decreases and the variability of the possible outcomes increases.

This can affect all kinds of things. Intelligence is certainly one of them, but other readouts of robust development include things like height and general physical health and mortality from many causes. The idea that intelligence is just one general fitness indicator is supported by the fact that it is correlated with these other factors. For some of these, like deaths from accidents, the correlation may really be driven by intelligence itself. But for others, like height, it is harder to see that kind of causal connection. It seems far more likely that both factors instead reflect some underlying, unmeasured parameter—developmental robustness.

As we noted in chapter 4, another factor that reflects developmental robustness is symmetry. The program of development encoded in the genome is played out largely independently on both sides of the body. Greater robustness should lead to a more consistent outcome and, thus, greater symmetry. Indeed, facial symmetry is correlated with judgments of attractiveness (along with many other parameters), consistent with evolutionary theory that suggests it is seen as an outward indicator of genetic fitness. And there is some evidence that symmetry also (weakly) correlates with intelligence—that higher IQ is associated with a greater degree of facial and bodily symmetry, on average. This even extends to the brain itself. There are a number of consistent structural differences between the left and right hemispheres of the brain, but there are also more random "fluctuating asymmetries" that can be measured from brain imaging. People with a greater degree of these random asymmetries tend to have lower general cognitive ability.

The robustness of the developmental program may also be very important clinically. As with the concept of cognitive reserve, where the robustness of neural circuitry determines vulnerability or resilience to secondary insults, we can think of "genomic reserve" in the same way. A greater general load of slightly deleterious variants might not have dramatic effects by itself (in clinical terms, at least), but could impair the ability of the developmental program to buffer the effects of other insults, especially new mutations. In this regard, it is interesting to note that lower IQ is, in epidemiological terms, a general risk factor for many psychiatric disorders. It is not just that the symptoms of the condition impact on cognition, which is often true, it is that having lower IQ prior to such symptoms predisposes to the condition. This is evident from the fact that the average IQ of *relatives* of people with psychiatric disorders is slightly lower, on average, than that of control individuals, even though those relatives are themselves clinically unaffected. We will see in chapter 10 how a general load of common risk variants may predispose to psychiatric conditions like schizophrenia, possibly by modifying the effects of more disease-specific rare mutations. It is at least plausible that this risk is mediated by general effects on genomic reserve and developmental robustness, which can also be indexed by IQ.

This brings us to one final wrinkle in thinking about factors that influence intelligence. It is clearly highly heritable, but not completely, leaving plenty of room for nongenetic effects. We have seen how environmental and experiential factors play crucial roles in determining people's absolute intelligence levels (though they may not affect relative rank as much, depending on how much variance there is in such factors). But there may also be a sizable contribution from intrinsic developmental variation—differences in the outcome of development from run to run. MZ twins who share both genetic variants and upbringing and schooling experiences tend to be highly similar to each other in IQ, but still differ somewhat. This suggests that intellectual potential may be even more innate than heritability estimates would indicate, depending not just on the developmental program encoded in the genome but also on the specific outcome of that program in each individual.

GENIUS

There is one area where developmental variation may play a very important role—that is in the really exceptional cases that we call "geniuses." There are many people with very high IQ scores in what is colloquially known as "the genius range" (though that term is not actually used in IQ testing), but I'm talking more about those few people with really unique intellects—the ones who can somehow see and grasp concepts that most mortals cannot, even their highly educated and accomplished peers. As Arthur Schopenhauer is supposed to have said: "Talent hits a target no one else can hit; Genius hits a target no one else can see."[1]

---

[1] Quoted in Gregory Bergman, *The Little Book of Bathroom Philosophy* (Gloucester, MA: Fair Winds, 2004), 137.

Intellectual, as opposed to creative, genius is most easily and widely recognized in the fields of physics and mathematics, the purest expressions of abstract intellect. (That may get a few philosophers' backs up!) Names like Newton, Leibniz, Einstein, Gauss, Ramanujan, Curie, von Neumann, and Feynman spring to mind. This is speculative, but one could certainly argue that the intellects of these remarkable individuals were not just at the extreme end of the quantitative distribution, but somehow worked qualitatively differently.

We know remarkably little about what sets such minds apart. It is not really clear that what we have learned about the underpinnings of intelligence across the general range even applies to such instances of true genius. Twin and family studies have shown that inheritance patterns at the very high end of the IQ distribution do not differ from the rest of the distribution—people with very high IQs tend to have relatives with very high IQs. But that extreme end of the normal distribution may not really define truly exceptional minds—the qualitative differences in intellect that set true geniuses apart can probably not even be captured by standard IQ tests.

There is little evidence to tell us whether true genius of that kind is genetic at all. An early study by Francis Galton, whom we met earlier, purported to look at *Hereditary Genius*, but Galton's test case was, rather immodestly, his own extended family! The term genius might well be applicable to Galton himself—he certainly had an unusual, expansive mind and a knack for seeing things that no one else had even conceived of before. And his family did contain many eminent and accomplished individuals, most notably his half cousin Charles Darwin, as well as their grandfather Erasmus Darwin and several members of the Wedgwood clan of inventors and industrialists. These were certainly very clever and productive people, but, of course, this concentration of familial success cannot be separated from their wealth and opportunity. It is ironic that Galton, who developed twin studies for just that purpose of distinguishing effects of shared genes from shared environment, did not feel the need to apply it in the case of his own family.

For the most part, though, the families of known geniuses have tended to be rather unexceptional. There is nothing in most of them to suggest that the relatives of geniuses such as those mentioned above were of particularly high IQ, at least not as a general rule. There are, however, ways in which an effect can be genetic but not obviously familial. One is if it is due to the effect of de novo mutations. Autism is often due to such mutations occurring in the sperm or eggs of parents, and is sometimes associated with "savant" abilities—particular isolated talents for lightning calculation or prodigious feats of memory. We saw in chapter 7 that such savant skills may reflect a combination of autism and the cross-perceptual condition of synesthesia, such as number-space synesthesia, where numbers are perceived in particular positions in space. This may allow a qualitatively different way of manipulating numbers from that employed in traditional arithmetic. However, while many geniuses in mathematics and physics may have had some autistic or synesthetic traits, those people recognized as autistic savants do not usually show the creative, innovative intellects of true geniuses. In fact, many are diagnosed with frank intellectual disability.

Another theory proposes that genius emerges from a qualitative change associated with particular *combinations* of genetic variants, which, when separated from each other in relatives, do not have such potent effects. That idea is possible, and certainly those kinds of nonlinear, nonadditive interactions between multiple genetic variants can indeed lead to larger differences than one would expect from the simple sum of the effects of the individual variants involved. Such nonadditive effects may be especially important at the extremes of quantitative traits.

Unfortunately, that theory is almost untestable. What we would need to know is whether the identical twins of such exceptional people also showed the same kind of genius. Would Alfred Einstein have been as insightful and intellectually creative as his twin Albert? Would James von Neumann have matched his twin John's accomplishments? We'll never know because Alfred and James did not exist. The genetic theory suggests they would have been, but another alternative is that the brains of Albert and John developed the way they did more by chance than genomic design. Any developing dynamic system that has nonlinear interactions will sometimes, very rarely, show transitions into qualitatively distinct states due simply to noise in the system. Most of the time the noise across multiple components and subsystems will cancel out, but there can be very rare occasions that arise, simply by chance, where the noise pushes some combination of parameters into a configuration that leads to a quite distinct outcome. Regrettably, that idea is also effectively untestable and, unless human cloning really takes off, we may never know if there is any truth in it.

CHAPTER 9

LADIES AND GENTLEMEN, BOYS AND GIRLS

Are men and women really that different? Obviously, physically they are, but behaviorally, psychologically? Well, if they weren't, a lot of stand-up comedians would have to go looking for new material. Men and women clearly do behave differently in many ways—on average, at least. The real question is how do they get that way? If we only consider humans in isolation, it can be extremely difficult to dissociate the influence of biological differences from those due to cultural norms and expectations—indeed, these two forces clearly interact in influencing patterns of behavior. But we did not spring, as a species, fully formed from the head of Zeus. We are evolved animals, with a genetic heritage honed over millions of years to ensure the survival of all our ancestors—hominid, primate, simian, and so on, back to the earliest animals. We can thus approach this question from a different angle, examining the biological basis of sexual differentiation and sexual behavior in other mammals, before considering the important effects of culture in humans.

And we can start with the most basic question of all: Why do we have sex? Not "have sex" in that sense—I mean why does sex, as in sexual reproduction, exist? It doesn't have to. It's quite possible to reproduce asexually—lots of organisms do it. We could be budding off little clones the whole time, without having to go to all the trouble of finding a mate. We could, but, besides being decidedly unromantic, there are several problems with asexual reproduction.

Foremost among those is that mutations accumulate over time in each clonal lineage. The only way to get rid of them is for individual lineages to die out. That works okay for small, rapidly dividing creatures like bacteria because they can produce so many individuals, but isn't a good strategy for larger organisms, where it takes a lot more resources to make a new individual.

In addition, because mixing of genetic material between individuals does not occur, it limits the generation of genetic diversity to the random sequences of mutations that arise in each lineage. This means a lot of the possible genetic "space"—all the possible combinations of genetic variants—remains unexplored, which lowers possibilities for adaptation. This lack of diversity also leaves whole clonal populations vulnerable to new infectious agents or to changes in the environment.

Sexual reproduction gets around those problems by mixing the DNA of two individuals every time a new individual is created. This isn't as easy as it sounds, however. You can't just smush two cells together. It requires some complicated machinery for one cell to fuse with another and for their genomes to be combined—machinery that we see in specialized germ cells. These cells are also special in that they each contain only *one copy* of each chromosome, instead of the normal two, so that when they fuse, the resultant organism will have two copies again.

But since you don't want two cells of one individual fusing with each other (i.e., self-fertilization, which would defeat the purpose), these germ cells come in two varieties, sperm and eggs. Sperm can't fuse with other sperm, and eggs can't fuse with other eggs. To keep them separated thus requires *two different sexes*—one that makes only sperm and one that makes only eggs.

Now that has some very interesting ramifications. In multicellular animals, it means that not only is there a difference between the sexes in the differentiation of the germline, there must also be a difference in the reproductive organs—the bits required for getting the sperm and eggs together. Typically, the sperm are the ones that travel, which for mammals means the fertilized egg develops inside the female. That requires a whole other set of anatomical and physiological specializations not needed in males. It also drastically changes the ecological roles that each sex plays.

Inevitably, this leads to behavioral differences between the sexes. The most obvious of these relate to mating itself. Having two sexes means that each individual can only successfully reproduce with a subset of the other individuals in the species. Because mating is energetically costly—both in expended effort and opportunity costs—and also dangerous if there are predators around, systems have evolved to enable recognition of appropriate partners. That means there must be something outwardly different between the two sexes that can be sensed: they must look or sound or smell different. And it means there must be some neural circuitry to detect those differences. And, finally, there must be some mechanism that says what to do with that information—some neural basis for sexual preference. The most efficient solution for that need, and the one that evolution has settled on, is to prewire that preference into the brains of each sex.

Of course, an animal doesn't necessarily want to mate with just *any* member of the opposite sex. To give its own genes the best chance to survive and be passed on, it wants to combine them with other genes that don't carry lots of mutations, so that the offspring are as healthy as possible. Raising young involves a huge investment, so choosing the right mate becomes a crucial decision. In mammals, this is much more true for females, because they make a much larger investment in their young. They have to carry the fetus, which is energetically costly and dangerous, and which also eliminates any further mating opportunities during that time. Males, on the other hand, can go off and inseminate another female as soon as the opportunity arises.

Also, infant mammals have to be nursed, which only females can do, meaning females invest more in the care of their offspring—indeed, in many species, males play no part at all following mating. This means the number of males ready to breed is usually much higher than the number of females, many of whom are either pregnant or already rearing young. In monogamous species, males stick around and contribute to rearing offspring, but even under those circumstances, their investment is lower than that of the females and their options for other matings higher. For all those reasons, it makes evolutionary sense for females to be much choosier in selecting mates than males, and for males to compete for these opportunities. This kind of sexual selection, which acts as a quality check *before* investing the resources to actually make offspring, can dramatically drive further differentiation of the sexes, both physically, in anatomy and physiology, and behaviorally (which really means in neuroanatomy and neurophysiology).

SEXUAL SELECTION—MAKING THE SEXES DIFFERENT

As first noted by Charles Darwin, sexual selection can act like an escalating arms race, driving some truly bizarre adaptations and behaviors. If females are choosy, hoping to select mates with higher evolutionary fitness, then males become competitive, aiming to show off their relative fitness with everything from ornate and energetically costly displays, like the peacock's tail, to the more direct route of simply knocking lumps out of each other. This can lead to differences between the sexes in all kinds of nonreproductive behaviors, especially including aggression and violence. Such differences can also be driven by a division of labor within species, with sex differences in ecological roles, such as nurturing offspring, hunting, foraging, defending territory, social grooming, and other activities.

In primates, the lineage from which humans emerged, these forces have led to males having significantly greater body mass, muscle mass, and bone thickness and density. In species that fight with fangs, the canine teeth are also commonly much larger in males. The extent of these differences varies a lot, however, in ways that reflect the mating and child-rearing strategies and social organization in each species. Those that mate for life and in which both parents invest in rearing offspring, like gibbons, tend to have lower levels of sexual dimorphism (physical differences between the sexes in morphology), while those that continually compete for mates—especially for harems of females, as in gorillas—can show enormous differences between males and females.

Early hominid species also showed substantial sexual dimorphism, as do modern humans. Human males are 15%–20% heavier than females, but have about 40% more muscle mass. Men also have thicker skulls, especially in the front, which may reflect the fact that we like to punch each other in the face a lot. Other differences, such as facial hair, could be due to sexual selection (women may find beards sexy), or may act as dominance displays in competition with other men (to avoid ever getting to the punching each other in the face part). Human males are also much more physically aggressive than human females, as we will discuss below.

Sexual selection affects females too, of course—they also compete with each other to attract the best mates and to encourage male fidelity and investment in offspring. This can drive exaggeration of indicators of fitness and fertility. Since females' reproductive ability declines with age, these indicators include retention of more juvenile characteristics such as more delicate facial features, higher-pitched voices, and reduced body hair. They also include a greater percentage of body fat and its selective distribution on hips, breasts, and buttocks. These fat deposits emerge with sexual maturity and are needed for ovulation, pregnancy, and lactation. They may thus signal fertility, in turn driving a male preference for these traits.

All of these observations are indicative of sexual selection having played an important part in our evolutionary lineage, including in our early human ancestors. These forces have driven differences not just in our bodies but also in our brains and our patterns of behavior. It is thus not only plausible that such innate sex differences would exist in humans (and demonstrable that they do), it is completely implausible that they would not. It would take a particularly virulent form of human exceptionalism to expect that we should differ from every other species of mammal in this regard.

The question is what kinds of traits do these sex differences affect? Sexual preference is the most obvious one—so obvious that we take it for granted, as if it requires no explanation—but we will also explore below effects on aggression, personality traits, interests, cognitive traits, and even sizable differences in susceptibility to neurological and psychiatric disorders. All of these differences have a physical basis. The brains of males and females are literally wired differently, both in neuroanatomy and in neurochemistry. From studies in humans and, especially, in other animals, we now know a lot about how they get that way.

WIRING MALE AND FEMALE BRAINS

Sex determination in mammals starts with the X and Y chromosomes. In addition to 22 other pairs of chromosomes in each cell (called autosomes), mammals also have either 2 X chromosomes, if they are female, or an X and a Y if they are male. The X chromosome is quite large and carries about 2,000 genes spaced out along its length. These are involved in all kinds of functions, just like genes on the autosomes. The Y chromosome is a very different beast. It's tiny, by comparison, with only about 200 genes, and most (but not all) of these are involved in male-specific functions, especially in making sperm.

Germ cells—sperm and eggs—have only half the genetic complement of normal cells. That is, they carry only a single copy of the genome, while other cells have two copies. In females, that means a single copy of the X chromosome goes into each egg. But in males, either a copy of the X or a copy of the Y gets put into sperm. Fertilized eggs will therefore either inherit two copies of the X or one X and one Y. So, imagine you're a little fertilized egg (as you were once), sitting there about to develop and trying to figure out if you should turn into a male or a female. You don't want to get this wrong and you can't be indecisive about it, either. That difference in the X and Y chromosomes is all you've got to go on, but it's enough to switch development down one route or the other.

Sexual differentiation proceeds in two stages. First, once formed in the early embryo, the cells of the initially indifferent gonads are directly affected by the presence or absence of the Y chromosome—if it's there, they develop as testes, if not they develop as ovaries. This depends entirely on the presence of one specific gene on the Y chromosome, known as *SRY*. If this gene is not functional then XY animals (mice, in this case) will develop as females. And if the gene is put somewhere else in the genome—on one of the autosomes—then XX animals that have it will develop as males. No other genes are required to initiate this switch, though many other genes are involved in the subsequent sexual differentiation. These other genes are present in both males and females—they are just regulated differently. The protein encoded by the *SRY* gene acts as a transcription factor—it regulates the expression of other genes in the cells of the gonads. In males, it switches on a cascade of gene expression that causes the gonads to differentiate as testes. In females, this doesn't happen and the female pattern of gene expression is turned on instead, leading to the differentiation of ovaries (see figure 9.1).

Male or female differentiation of the gonads is the primary event in sexual differentiation, but it is followed by the release of male or female hormones from the gonads. In particular, the testes start to produce testosterone and that leads to the secondary sexual differentiation of the rest of the body, including the brain. (Incidentally, many other species

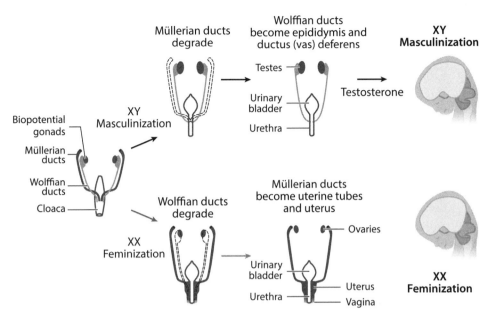

Figure 9.1 Sex determination in mammals. The initially undifferentiated gonads develop as testes if the Y chromosome is present, and as ovaries otherwise. The testes then produce testosterone, which masculinizes the developing brain. The X and Y chromosomes also contribute directly to brain masculinization and feminization.

use a different mechanism, where it is the number of copies of the X chromosome that initiates the sexual differentiation pathways, independently in each individual cell in the animal, with no role of the Y chromosome and no influence of hormones—more on that later.)

The crucial role of sex hormones has long been appreciated, but exactly how they function took some working out. First, they play two different roles. In adults, they have an *activational* role: hormones like estrogen and progesterone in females and testosterone in males are involved in regulating all kinds of reproductive behaviors, most notably the estrous cycle in women. Testosterone is also involved in male puberty, and, as is obvious from doping scandals in sports, higher levels lead to increased capacity to build muscle. Both male and female sex hormones also have acute effects on behavior in adult mammals, including humans, especially on sexual drive and receptivity.

But, crucially, those responses differ between males and females exposed to the sex hormones. There is already some underlying difference between male and female brains, which relies on an earlier *organizational* function of the sex hormones. Studies in many species have shown that this takes place during an early critical period of brain development, when the brain gets either masculinized or feminized.

In rodents, there is a surge of testosterone produced around or just after birth. To test whether this had any permanent effects on brain development, male rats were castrated at that age and their later behavior analyzed. Even when testosterone was administered to them as adults, these male rats displayed a much lower tendency to mount female rats and attempt to mate with them. Conversely, when female rats were injected with testosterone in the first week of life, they later showed male-like tendencies to mount other females and were much less receptive to being mounted by males. Remarkably, if either the castration of males or the administration of testosterone to females was done later in life, after the first week, these permanent effects were not observed. Similar effects have been seen in guinea pigs, monkeys, and other species. This emphasizes the importance of this early critical period of brain development, when it can be either masculinized or feminized.

But there was a surprise coming. When the researchers injected the young female rats with *estrogen* instead, their behavior was also masculinized—in fact, it was even more effective than testosterone. That didn't seem to make much sense, and the explanation for it lies in some rather arcane biochemistry of these hormones and the proteins that interact with them. Female fetuses don't normally make high levels of estrogen before or around birth, and what they do make is mostly bound up by a protein called alpha-fetoprotein, which prevents it from entering the developing brain. Testosterone, on the other hand, is made at high levels by male neonates and does enter the brain. Surprisingly, though, when it gets there it is mostly chemically converted to estrogen. This is done by an enzyme called aromatase, which is expressed at high levels in the brain.

It turns out that it is estrogen that does most of the job of masculinizing the developing male brain in rodents. This was shown by looking at mice that have mutations in genes encoding the aromatase enzyme or the two proteins that act as estrogen receptors. It's their job to detect estrogen in a cell and turn on or off various genes in response to it (or make some other biochemical changes in the cell). If testosterone really acts only after being converted to estrogen, then mice with mutations in these genes should show defects in masculinization of the brain. This is exactly what is observed—male sexual behavior is effectively abolished by these mutations. They also show increased female sexual behavior (increased receptivity to being mounted) and decreased aggression (male mice are usually vastly more aggressive than females).

The effects of testosterone (via the estrogen pathway) are also directly visible on brain structure. The brains of male and female rodents don't look overtly different on a gross level, but on a finer level there are many sex differences. For example, there is a particular area of the hypothalamus that is about five times larger in males than in females. In fact, it is named the "sexually dimorphic nucleus of the pre-optic area" (SDN-POA). (In humans, where it is also sexually dimorphic, it is called the interstitial nucleus of the anterior hypothalamus-3, or INAH-3; see figure 9.2). The hypothalamus is a brain region with many subdivisions that controls many aspects of physiology and behavior, including release of hormones and reproductive behaviors in adults.

The SDN-POA is bigger in males due to differences in cell death—many of the cells that would contribute to this region die in females, but are protected from this fate in males. This is clearly due to the action of testosterone (again, via estrogen) because the SDN-POA is female sized in castrated males and male sized in females given testosterone during the critical first week after birth. There are many other regions that also show differences in size (really, cell number) between males and females, though not to quite the same extent. Many of these show the opposite pattern, however, being larger in females than in males.

There are also sex differences in the number and density of connections between different brain areas. These are independent of the differences in cell number and reflect additional effects of hormones on genes that promote the growth of nerve fibers between various structures. The brains of male and female mammals are thus literally wired differently. This extends to an even finer level, when looking at the distribution of synaptic connections. In many brain areas, including various regions of the hypothalamus, nerve cells have a different number of branches and synaptic connections with other cells in males versus females. In some regions, including the hippocampus (a higher brain area involved

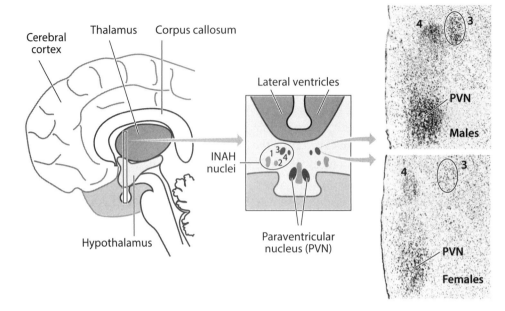

Figure 9.2 Sexual dimorphism in the hypothalamus. There are far fewer cells in the INAH-3 nucleus (*circled*) of the hypothalamus in females, compared with males. (Reproduced from R. A. Gorski, "Hormone-Induced Sex Differences in Hypothalamic Structure," *Bull. Tokyo Metrop. Inst. Neurosci.* 16, suppl. 3 (1988): 67–90.)

in memory and many other functions), the number of synapses also fluctuates in adult females with the estrous cycle.

If we go even deeper, to the level of biochemistry and gene expression, we also see many differences between the cells in male and female brains. Most of these are caused secondarily through the influence of hormones, but some of them are directly caused by differences in the sex chromosomes. We tend to think of sex at the level of a whole organism, but each cell has a sex and male cells are intrinsically different from female cells in many ways. Even before hormones are produced by the gonads (before the gonads even differentiate, in fact), there are already some large differences in gene expression between the cells in male and female brains.

These arise directly from the fact that the individual cells each have either two X chromosomes or one X and one Y. In order to make sure gene expression is generally equal between males and females, one of the X chromosomes, at random, is normally "shut down" in female cells. However, not all of the X chromosome is shut down—some genes "escape" this inactivation. As a result, females have two active copies of some X chromosome genes, while males have only one. Some Y chromosome genes are also expressed in the brain during development, obviously only in males. These initial differences in a few genes affect the expression of other genes so that the overall gene expression profile differs substantially between male and female cells. These differences can affect lots of biochemical parameters—even, for example, how sensitive these cells are in a dish to various drugs. In humans, dozens to hundreds of genes (on the autosomes) appear differentially expressed in the brain between males and females. In rodents, these direct effects of the sex chromosomes on gene expression in the brain have been shown to also contribute, as well as hormonal influences, to sex differences in various behaviors including aggression, parenting, social interactions, and even pain perception.

Sexual differentiation of the mammalian brain thus involves a complicated and coordinated genetic program, or really *a switch between two alternative programs*. These two programs drive divergent developmental trajectories in male and female brains, which result not just in structural differences in adults but also differences in expression of many proteins that regulate neural function and plasticity. These include the receptors for the sex hormones themselves and also for neuropeptides like vasopressin and oxytocin, which regulate pair-bonding and parental behaviors differentially in males and females. There are also marked sex differences in the degree of synaptic plasticity—how synaptic connections are modified by experience—and especially its sensitivity to sex hormones or stress hormones. Not only do male and female brains come prewired differently—they also change differently.

This general scheme is very highly conserved across many species, though the precise details can vary. In primates, including humans, testosterone seems to act more directly through its own receptor, called the androgen receptor, to masculinize the brain, rather than through conversion to estrogen. Unlike in mice, mutations in the aromatase gene in humans do not seem to lead to a change in sexual differentiation—males with such mutations are attracted to females at normal rates. This suggests that the conversion of testosterone to estrogen is not so important in developing human brains.

By contrast, mutations in the androgen receptor lead to androgen insensitivity syndrome, where XY individuals develop as females, despite the presence of testes and the production of testosterone. The cells that normally would respond to testosterone cannot detect it without a functional androgen receptor. This mutation apparently also precludes masculinization of the brain, as the XY women with this condition are attracted to males at rates equivalent to XX women. Conversely, women exposed to higher than normal levels of testosterone during development—in a condition called congenital adrenal hyperplasia, in which the adrenal glands produce high levels of steroids including testosterone—show increased rates of male-typical behavior as children, increased feelings of male gender identity as adults, and increased rates of sexual attraction toward women.

Despite this difference, the overall plan in humans seems to be the same as in other mammals—that is, sex hormones have an early organizational role during a critical period, which prewires the male and female brain differently, and they also have a later activational role, involved in driving or regulating reproductive behaviors in adults (with the responses in adults being dependent on the initial organizational effects). While we can't see differences at the cellular level in human brains (not while they're being used, anyway), we can see differences at the gross structural level using neuroimaging.

SEX DIFFERENCES IN HUMAN BRAINS

The most obvious difference between male and female brains is that male brains are larger, by ~10%. This is only partly explained by overall larger body size—that is, males' brains are even bigger than you'd predict based on their increased body size. But there are also many more specific differences in particular regions or nerve tracts, or in the overall organization of neuronal networks.

It should be emphasized that all of these are *group average differences*—for all of these parameters there is a distribution of values within males and a distribution within females, and, even though the mean values of these distributions differ between males and females, the distributions still overlap substantially. This is the same situation as with height. Males are taller than females, on average, but it is not the case that every male is taller than every female. The sex effect is only one (relatively minor) influence on variance in height, most of which is due to general genetic variation or stochastic developmental variation. The group difference alone doesn't have any predictive power about the trait in individuals—you can't say how tall someone is just from knowing whether the person is male or female. But what we can infer is that if any given woman were otherwise genetically identical but were male, she would probably be a bit taller than she actually is. And if any male were female instead he would probably be a bit shorter. The sex effects simply add on top of the underlying variation. The same logic applies for differences in brain structure, at least at the gross level assessed by neuroimaging.

It is important to note that many of the historically reported findings in this field have been inconsistent, replicating in some subsequent studies but not others. This is most likely due to small sample sizes, usually involving only tens of subjects, which may have given spurious results just by chance. If a sex difference is small relative to other sources of variation in any given structure, it should take large samples to reliably detect it. Recently, much larger studies have been carried out, comparing many hundreds or thousands of subjects, which have found clear sex differences in many different brain regions.

There are, first of all, differences in the size of various lower brain structures, including parts of the thalamus and basal ganglia (the striatum and pallidum, for those keeping score), between males and females. Of course, there may be many other sex differences that are too subtle to see on a neuroimaging scan. These include, for example, INAH-3, the human equivalent of the SDN-POA region of the hypothalamus, mentioned above, which is over two times larger in men than in women. Another area also involved in sexual behaviors is the bed nucleus of the stria terminalis (BNST); the central part of the BNST is also about twice as big in males as in females. These regions are too small to measure in neuroimaging scans—they require sectioning and microscopic analyses of postmortem brains to detect.

In the cortex, where more attention has been focused, there are also numerous areas that show sex differences by neuroimaging in large cohorts—some larger in females, others larger in males. These differences are small, relative to the range of variation seen within either sex, so that the distributions of sizes of any particular area overlap substantially between males and females. In a study by psychologist Daphna Joel and colleagues looking at the 10 most dimorphic areas, very few individuals were at either the extreme male end or the extreme female end for all 10. This led the researchers to conclude that individuals were a "mosaic" of masculinized and feminized regions and we really shouldn't be thinking of "male" or "female" brains at all. It is, however, exactly the pattern one would expect if the sex effect is small and laid on top of independent variation in the sizes of each of these regions arising from other factors.

In fact, a very similar situation holds for facial morphology. Male and female faces differ on many different parameters—nose length and shape, brow ridges, face width, jaw size, chin shape, and so on—each of which also has its own underlying variation, largely independent of the others (see figure 9.3). If you just look at one of these, you can't tell a person's sex well at all, because the distributions overlap so much. But if you take many of them into account at the same time then you can tell the difference between male and female faces with over 95% accuracy— human beings are really good at that, and now so are facial recognition programs. The same is true for the brain imaging data—a "multivariate" classifier that considered the size of all 10 brain regions at once from the same brain scan data was able to classify males and females with over 90% accuracy. There thus clearly are male and female brains in the same way that there are male and female faces.

While it's possible to detect differences in the size of various brain regions, this is an extremely crude way to try to assess how the brain works. You could, first of all, find no differences with neuroimaging but still have many differences in microstructure, neurophysiology, and gene expression, which could profoundly influence function. Moreover, whether some region is bigger in men or women (or not) doesn't really tell you much about whether it works differently. What's more interesting is how each region is connected. At a structural level, this can be analyzed by looking at the network of nerve fibers connecting each region. A study of nearly 1,000 people found strong differences in the network architecture between males and females.

The main trends were that males tended to show greater local connectivity, with more tightly defined clusters of interconnected points,

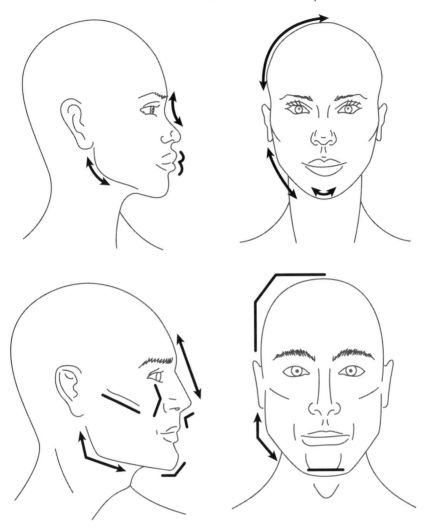

Figure 9.3 Sex differences in facial morphology. Male and female faces differ, on average, in multiple properties, including bridge of the nose, brow ridge, square versus rounded jawline and chin, eyebrow curvature, thickness of the upper lip, and others. No one of these variables is diagnostic of sex, but collectively they allow very accurate classification.

while females showed greater connectivity between clusters. (This finding has been replicated in two other studies.) Females also tended to show greater connectivity between the two cerebral hemispheres, while males tended to show greater connectivity within each hemisphere. These data strongly support claims that females have a *relatively* larger corpus callosum—the thick band of nerve fibers connecting the two hemispheres—than males. This finding had been inconsistently replicated across many small studies, leaving its generalizability unclear, but the results of this very large study seem conclusive.

Now, you might say that observations of differences between the brains of *adult* males and females don't prove that they actually started out different, and that is perfectly true. Perhaps those differences arise over time through the mechanisms of brain plasticity, due to the fact that males and females have different experiences and are treated differently, by their parents, their peers, and society in general. In isolation, that is at least conceivable, though we have seen that adult brain structure in general is actually highly heritable, reflecting continued, strong genetic influences on growth and maturation after birth. Moreover, the fact that similar differences are seen in every other mammalian species and drive evolutionarily important behaviors makes the argument that they are purely culturally driven in humans much less likely. In essence, it demands two extra things: first, our evolutionary heritage of sex differences as mammals and primates would have to have been wiped clean, despite those differences being highly adaptive, and, second, cultural practices would have to have arisen that effectively re-create the same outcome.

More direct evidence for innate differences comes from imaging studies of infants and children, which already show clear sex differences in brain structure. When scanned within a couple of weeks of birth, males already have a larger overall brain (~6%–9% larger) and greater volume of both gray and white matter, even after adjusting for birth weight. This is true even though the brain at that age is only 35% of adult size. Regionally specific sex differences have also been observed in neonates, with some areas larger in males but others larger in females. Male and female brains thus start out different at birth, even before they grow to full size. The brains of boys and girls also grow at dramatically different rates, with girls' brains maturing much earlier than boys', reaching peak size for various structures two to four years earlier. Some of the particular differences observed in neonates prefigure those seen in adults, while others disappear as brains mature and new differences also emerge, especially with puberty.

Humans are thus like all other mammals in having discrete mechanisms of sexual differentiation that include the brain. These lead to differences in structure at a macroscopic level and much more subtle (but possibly more consequential) differences at the cellular level, for example in neuronal branching or synaptic connectivity, as well as differences in neurochemistry and gene expression that fundamentally alter important neural functions, such as synaptic plasticity.

The important question is what do all these differences in structure *mean*? Many attempts have been made to link the size of various regions or differences in various network parameters with particular behavioral differences between men and women, but no particularly solid relationships have emerged. (There aren't, in fact, many solid relationships between the size of particular brain structures and specific behavior across people generally.) It's natural enough to assume that if there is a structural difference in the brain, it is most likely contributing to some difference at the behavioral level.

There is an alternative way to think about it, however, which is that some of those structural differences may actually be a means of *compensating for* differences in physiology between men and women or differences in other parts of the brain, so as to keep them functioning as similarly as possible. Evolution has a tricky job to do—it has to make male and female brains different enough to drive appropriate sex-specific behaviors, without impairing general behaviors required for survival of both sexes, which may involve many of the same brain regions. It is probably far too simplistic to expect structural differences of single regions to correspond to differences in psychological traits or behaviors, when they are embedded in networks with many other distributed differences, some of which likely counteract each other. However, even if we cannot currently associate specific brain differences with specific behavioral differences, we can at least see what sorts of behaviors differ between men and women generally.

So, are men and women really that different in psychological and behavioral traits? In a word, yes. And no. In domains such as cognition or personality there are many consistent and interesting group average sex differences, but these are small in magnitude and the distributions of the individual traits involved are highly overlapping. When looked at collectively, however, the combined profiles of multiple traits do show larger sex differences, as we will see. At a clinical level, there are very large differences in the rates of many psychiatric and neurological disorders between men and women—understanding the basis of these differences is thus crucially important. But the most obvious psychological differences between males and females are in exactly the traits you would predict—those affecting sexual behavior and reproduction.

SEXUAL PREFERENCE AND SEXUAL ORIENTATION

It is hard to think of a more clearly and strongly genetic trait in humans (or any animals) than sexual preference. It is so commonplace that we take it for granted—that males are mostly attracted to females and females are mostly attracted to males, like that just happens. But if you just think about *humans*, you can see how this isn't a simple default position—it is actually two different states, each of which requires active processes to establish. The vast majority of humans who inherit a Y chromosome are primarily attracted to females, while the vast majority of those who do not are primarily attracted to males. That is an incredibly potent genetic effect.

It is also clear that sexual preference is strong and effectively innate, even though actual sexual interest does not emerge until after puberty and sexual maturation. This is, first of all, common experience—that we did not choose our sexual preference, that we are not free to change it, and that it did not have to be learned. It is also supported by the comprehensive failure of many different methods that have been used to try to get people to change their sexual preference. There are, in addition, a number of conditions affecting sex hormone pathways in humans that alter sexual preference in ways consistent with effects in other mammals on the early development of the brain. Sexual preference is not just prewired into our brains, it is remarkably hardwired.

The pioneering work by Alfred Kinsey and colleagues in the 1940s suggested that variation in sexual preference lay along a continuum—that people varied in how strongly and exclusively they were attracted to one sex or the other.[1] More recent work argues strongly against this view

---

[1] A. C. Kinsey, B. P. Wardell, and E. M. Clyde, *Sexual Behavior in the Human Male* (Philadelphia, PA: Saunders, 1948); A. Kinsey, W. Pomeroy, C. Martin, and P. Gebhard, *Sexual Behavior in the Human Female* (Philadelphia, PA: Saunders, 1953).

and indicates instead that sexual preference is much more categorical, for both heterosexuals and homosexuals. Statistical methods looking at the underlying structure of variation in responses to questions about sexual attraction, identity, and experiences clearly separate people into two main groups (either attracted to males or attracted to females, regardless of their own sex), with very few identifying as or classified as bisexual (higher among women than among men). This is consistent with the picture described above—that sexual differentiation operates by switching between two alternative, even competing, states.

Given the complexity of the genetic programs regulating masculinization or feminization of the brain, it's not surprising that there can be genetic or developmental variation that affects the outcome. When that happens, you can end up with a mismatch between a person's sex (defined by sex chromosomes and gonadal development) and that person's sexual preference. There is a tendency to think of "sexual orientation" as a single trait—either heterosexual or homosexual. But in fact it really describes the intersection of two different traits—gonadal sex (male or female) and sexual preference (toward males or females). There is strong evidence that homosexual orientation is just as innate as heterosexual orientation. It is also partly genetic—but only partly.

Twin studies show a clear genetic effect. A variety of studies have found that when one of a pair of MZ twins is homosexual, the other one also is around 30%–50% of the time. That rate in DZ twins is only 10%–20% (still a lot higher than the overall population rate). As for other traits, that difference in concordance between MZ and DZ twins indicates a genetic effect, with the heritability estimated at around 40%–50% (slightly lower for women than for men). Another important observation is that homosexuality in one DZ twin does not increase the rate of homosexuality in the other twin when they are of the opposite sex. This reinforces the view that masculinization and feminization of the brain are two different active processes involving two different sets of genes (i.e., heterosexuality is not a default state; it is two different states). Mutations that affect one pathway for the most part do not affect the other.

At present, we do not know the identity of most of the genetic variants that contribute to variance in sexual orientation. There are a couple of exceptions, however—for example, mutations in a number of genes encoding enzymes in the steroid biosynthetic pathways can lead to excessive production of sex steroids from the adrenal glands and cause congenital adrenal hyperplasia. As described above, women with this condition are exposed to very high levels of testosterone and have high rates of homosexual orientation and male gender identity as adults. Beyond those few genes with very rare mutations, we not only have not identified others, we also don't really understand the genetic architecture of the trait—we don't know how many distinct genes are involved, if it is caused by single mutations or multiple ones in combination, if these genetic variants are rare or common in the population, if they tend to be inherited or arise as new mutations in sperm or eggs, or if they persist in the population or are selected against. None of that ignorance argues against the major conclusion that sexual orientation is partly heritable, however—it just means we don't yet know the molecular details.

A small number of studies have looked for differences in brain structure or function between heterosexual and homosexual men or women. One highly publicized study by Simon LeVay found in postmortem samples that the INAH-3 nucleus (the human equivalent of the SDN-POA in rodents) was much smaller in homosexual men than in heterosexual men, more similar to the size seen in females.[2] Another study, by Dick Swaab and colleagues, found that the size of the BNST in male-to-female transsexuals was typical of females, suggesting a biological correlate in degree of masculinization of the brain with male gender identity.[3] Such studies are very rare, as they rely on postmortem samples, and these specific findings have not been independently replicated. They do fit, however, with observations in animals described above, where alterations to hormonal signaling at early stages lead to similar correlated effects on brain structure (such as the size of the SDN-POA) and on sexual orientation.

The fact that sexual orientation is only *partly* genetic demands an explanation. The pathways that normally establish sexual preference in males and females seem completely genetic (dependent on the presence or absence of the Y chromosome), but somehow the exceptions to those rules are not. This suggests a probabilistic relationship between people's

[2] S. LeVay, "A Difference in Hypothalamic Structure between Heterosexual and Homosexual Men," *Science* 253, no. 5023 (1991): 1034–37.

[3] F. P. Kruijver, J. N. Zhou, C. W. Pool, M. A. Hofman, L. J. Gooren, and D. F. Swaab, "Male-to-Female Transsexuals Have Female Neuron Numbers in a Limbic Nucleus," *J. Clin. Endocrinol. Metab.* 85, no. 5 (2000): 2034–41.

genotype and their sexual orientation. Most males have effectively a 100% probability of developing such that they are attracted to females. (If we cloned the average heterosexual male 100 times, we'd expect 100 of the clones to also be attracted solely to females.) But some males, who carry some specific genetic variants affecting the masculinization of the brain, have a lower probability of that outcome—perhaps only 50%, if we go on the concordance rates in MZ twins. If you cloned a homosexual male 100 times, we might thus expect 50 of the clones to be homosexual and the rest to be heterosexual. You can imagine a similar scenario for females who carry genetic variants affecting feminization of the brain (or preventing masculinization).

So what explains the difference in outcome? There is a natural tendency to look for some explanatory factor in the environment or in the experiences of people that can account for this nongenetic variance. But as we have seen many times, the processes of development themselves are highly variable. Small, random fluctuations in cellular components—the kind of molecular noise that is happening all the time—can, at certain points in development, lead to large differences in outcome. This may be especially true for sexual differentiation due to the fact that it acts in a switch-like manner—the two possible pathways are not just distinct, they are likely directly antagonizing each other to establish either the male or female state. Once you start going down one or the other pathway, this decision may be rapidly reinforced and consolidated. This makes evolutionary sense—ambiguity in this decision would likely lead to a high burden of individuals with low numbers of offspring.

Viewed from this perspective, it is perfectly understandable that a trait like sexual orientation could be only partly genetic, but still completely innate. A person's genotype confers a certain *probability* of one outcome or the other, but the actual outcome that arises in the individual depends on how development happens to play out. In this way, it is quite like handedness, which is also only partly genetic but apparently completely innate and resistant to outside influence. We don't choose our sexual preference any more than we choose to be right- or left-handed.

While sexual preference is the most obvious and strongest behavioral difference between males and females, there are many other psychological domains that also show consistent sex differences. The clearest of these are in physical aggression.

AGGRESSION AND VIOLENCE

Men and boys engage in far more physical aggression and violence than do women and girls. This finding is extremely consistent across cultures and across age groups. In the United States, 80% of persons arrested for violent crime are male, while 90% of homicides are committed by men. This number is even higher worldwide, at 96%, according to a 2013 United Nations report.[4] Archaeological and historical data indicate that this male propensity for lethal or traumatic violence was possibly even more prevalent in ancient times. Males also make up the majority of victims of violence—worldwide, 78% of homicide victims are men (not including those killed in wars).

This should come as no surprise, given the evolutionary pressures associated with our species' mating practices. There is clear evidence for sexual selection acting substantially through direct physical contests between males—not just to influence female choice but also to exclude weaker males from mating opportunities. These adaptations include increased body mass, especially upper body muscle mass associated with physical strength (about 90% greater upper body strength); increased bone density; and even reinforcement of facial bones, likely evolved to help withstand impacts. As in many other species with similar mating pressures, especially all the other members of the great apes, human males come armed and armored and inclined to fight.

This inclination is evident from an early age—across all cultures, boys engage in rough-and-tumble play (play fighting, wrestling, hitting, chasing) about three to six times more than girls do. This pattern holds in many other species—chimpanzees, monkeys, even rats. Interestingly, these differences can be observed in very young infants in these species, at times when circulating hormones are very low. This suggests that organizational roles of hormones during brain development are key, rather than activational effects on mature circuits. Manipulation of hormone levels during the critical period, as described above, can dramatically affect aggressive behaviors in mice, rats, and monkeys. In humans, girls

---

[4]United Nations Office on Drugs and Crime, *Global Study on Homicide 2013*, sales no. 14.IV.1 (Vienna: United Nations, 2013).

with congenital adrenal hyperplasia also show higher levels of physical aggression than unaffected girls.

The most straightforward interpretation of these observations is that human males, like those of other species, are innately inclined to engage in play activities that act as rehearsals for adult contests. This does not, however, mean that culture plays no role in the sex differences in aggressive behavior. It is certainly true that parents, other adults, and even other children have different expectations of and reactions to aggressive behavior for boys and girls. Rough-and-tumble play is far more likely to be tolerated among boys and discouraged among girls, which may obviously reinforce these initial tendencies. On the other hand, parents likely spend quite a bit of time discouraging it among boys as well! Indeed, our legal and penal systems are heavily aimed at preventing and punishing male violence, not promoting it.

Whatever the societal attitudes, it seems far more likely that they reflect rather than cause the sex differences in aggressive behavior. While they may contribute to ultimate differences through a kind of positive feedback, they are not the ultimate *origin* of those differences. After all, cultural expectations don't come from nowhere. If boys weren't naturally more inclined to engage in rough-and-tumble play, where would this expectation have come from? And why would it be so ubiquitous? If it's because adult men are more violent than women, then why is that? It can't be simply because our culture expects them to be, because then you're back to the same question of why that is. The simplest explanation is that we expect boys to engage in play fighting more than girls because we consistently observe that they do. And we expect men to be more violent than women because men actually are more violent than women.

PERSONALITY TRAITS AND INTERESTS

Men and women often behave in different ways in the same situations, which are more or less typical of their sex, on average. For example, men tend to be more competitive and assertive in their social interactions, while women tend to be more cooperative and conciliatory. Of course, much of the reason for that has to do with societal gender roles, norms, and expectations. But there could also be underlying biological differences in personality traits between the sexes that influence these patterns of behavior in a consistent fashion (and that lead to those expectations in the first place).

Many, many studies have examined this question, in very large samples, across many different countries in the world. The upshot is that there are indeed average differences in numerous personality traits between men and women, which are highly consistent across studies and across countries. If we consider the Big Five personality traits, women tend to score considerably higher on Agreeableness and Neuroticism, and slightly higher on Conscientiousness, while men tend to score slightly higher on Openness to ideas. Extraversion does not show an obvious effect one way or the other, but when it is broken down into subfacets it emerges that men tend to score higher on assertiveness and sensation seeking, while women tend to score higher on sociability and gregariousness.

Another personality scheme maps people along two major dimensions of dominance and nurturance. On this scale, males tend to score considerably higher on dominance and females considerably higher on nurturance. A related trend is observed when using a 10-, 16-, or 30-factor model of personality traits, which are narrower components than the Big Five traits. These show more specific and sometimes larger sex differences, with women scoring higher in sensitivity, warmth, and apprehension, and males scoring higher in dominance and emotional stability.

Of course, all of these differences are group averages and they are all small relative to the overall variation in these traits, so that the distributions largely overlap. Knowing the score for any single trait gives hardly any predictive information on a person's sex. However, as with facial morphology traits or brain imaging measures, when multiple traits are considered at once, in a multivariate model, it is possible to classify people into males or females with a high degree of accuracy. Whereas the distributions on individual traits still show an overlap of 60%–90%, the multivariate distributions only show an overlap of 10%, suggesting a real distinction in overall profiles.

These multivariate classifiers define archetypal profiles of personality traits for males, described by psychologist Marco del Giudice as "more open-minded, assertive, risk-prone, tough-minded, cold-hearted, emotionally stable, utilitarian, and open to abstract ideas," and for females as "more nurturant, warm, altruistic, submissive, risk-averse, tender-minded, emotionally unstable, and open to feelings and aesthetic experiences."[5] Analyses of interests and values also paint a similar overall picture, with men consistently tending to score higher in interests for *things* and women tending to score higher for interests in *people*. An alternative scheme for the types of things that people value shows that men score higher on "theoretical values," while women score higher on "social values." Both the interests and values schemes map well onto traits of Systemizing and Empathizing, developed by Simon Baron-Cohen, which show the same trends, respectively.

Broadly speaking, these profiles fit with evolutionary models, especially in terms of the sex differences in dominance and nurturance. It is therefore highly plausible that they reflect real, biologically based differences that are adaptive and that have been programmed into the human genome by natural selection. However, even more so than some of the other traits we have considered, sex differences in personality traits may also be particularly sensitive to cultural norms and expectations. Indeed, some argue that cultural pressures are completely responsible for the observed personality differences (i.e., tendencies to behave in particular ways in various circumstances). Several lines of evidence argue against the idea that culture is the primary origin of such differences, however.

First, the observed differences are extremely consistent in direction across a very large range of countries, both in the developed and in the developing world, across Europe, North America, South America, Africa, Asia, and the Middle East. These encompass quite divergent cultures, yet the personality differences are largely the same. What does vary is the *magnitude* of these differences, though not, perhaps, in the way you might expect. If these differences in behavioral tendencies were culturally driven, one might expect that cultures with more rigid and traditional gender roles would show greater sex differences. In fact, the opposite is the case. It is the most developed nations, with the highest levels of gender equality in social and legal terms, that consistently show the greatest degree of difference in personality traits between men and women.

---

[5] M. Del Giudice, T. Booth, and P. Irwing, "The Distance between Mars and Venus: Measuring Global Sex Differences in Personality," *PLoS One* 7, no. 1 (2012): e29265.

That's very peculiar, but there is a plausible, if somewhat arcane, explanation. There is a theory, with considerable experimental evidence to support it from animal studies, that traits showing sexual dimorphism should be highly sensitive to environmental stressors. The idea is that extreme sexual characteristics are costly—that is, in fact, the purpose of many of them—conspicuous consumption of resources to advertise the high level of resources an individual has. In times of plenty, the full range of sexual dimorphism will be expressed, while in harsher times or environments, where resources are scarce, the range will be narrower (specifically, the sex in which the feature was exaggerated will move back closer to the phenotype of the other one).

This is the case with height in humans. If we look across the world, the most developed countries with the greatest level of nutrition and health care show the greatest dimorphism in height between the sexes. The full potential of those two genomic programs can be realized under such conditions. The same thing seems to hold for personality traits—in particular, males seem to express a more archetypal masculine profile under such conditions than they do under more challenging or restrictive conditions.

In addition, sex differences in temperament—the more basic forerunner of personality traits—are observed even in very young infants. Males tend to show higher levels of what is known as surgency (high levels of activity and positive emotion, impulsivity, and engagement with their environment) from birth and are even, on average, more active in the womb. Female infants tend to show higher levels of effortful control, involving both greater inhibitory control and higher attention. These differences in temperament correspond only partly to differences in adult traits, but their existence does at least indicate that males and females are already behaviorally different, *even before birth* and prior to any possible enculturation.

Even if particular sex differences only emerge over time, that does not mean they are not genetically directed. The programs of sexual maturation encoded in the genome play out over many years and many psychological traits change dramatically with maturation (most notably sexual interest itself). From an evolutionary perspective, we can expect sex differences to be strongest after sexual maturation, when they become ecologically relevant, with the exception of traits in children that act as precursors to adult behavior.

An evolutionary, biological basis for sex differences in personality traits is also supported by the fact that similar differences can be observed in animals, like chimpanzees, where males are similarly higher in dominance and females higher in nurturance—again, even from an early age. This is apparent in social interactions and patterns of play that strongly resemble those seen in human infants (but obviously without a cultural explanation).

This is not to say that biological differences in personality between men and women are not amplified and reinforced by culture—they very clearly are, or at least they lead to cultural expectations. The archetypes described above are only average profiles to which any given man or woman will correspond more or less closely overall, and more or less closely on each specific attribute. These archetypes (which represent "very typical *examples*" of a man or woman) undoubtedly form the basis of stereotypes—fixed but oversimplified ideas of how every man or woman should be. The fact that it is unfair and prejudicial to apply these stereotypes to individuals does not, however, mean the archetypes are not accurate, as such. How we deal with and apply that information is a separate matter—clearly a hugely important one, but one that does not speak to the validity of the findings themselves.

COGNITIVE TRAITS

Despite the difference in brain size between the sexes, with male brains being on average 10% larger than female brains, there is no difference in average IQ scores between the sexes. This is a bit surprising, because there is a correlation between brain size and IQ generally, of about 0.40 (which holds *within* each sex). The fact that there is no sex difference in mean IQ suggests something really interesting—that there may be other differences in brain structure or organization between male and female brains that offset the simple effect of size on IQ. This reinforces the notion that some differences between male and female brains may be compensatory in nature, aimed at minimizing differences in overall function.

However, although there is no sex difference in mean IQ, there is a difference in the variance (see figure 9.4). The distribution of males is flatter than that of females, meaning there are fewer males near the

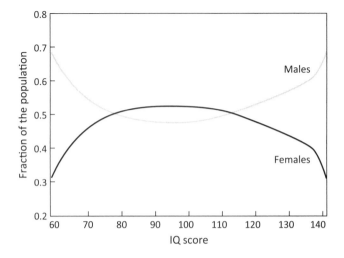

Figure 9.4 IQ distribution in males and females. Fraction of males and females at each point along the IQ distribution. There are more females near the center, and more males at both the low and high extremes. (Summarizing data from Scottish Mental Surveys as presented in W. Johnson, A. Carothers, and I. J. Deary, "Sex Differences in Variability in General Intelligence: A New Look at the Old Question," *Perspect. Psychol. Sci.* 3, no. 6 (2008): 518–31.)

middle and more at both the low and high extremes. The distribution of IQ across the whole population is not really "normal"—that is, not a perfect bell-shaped curve. Instead, it is quite skewed toward the low end (see figure 8.2). This can be attributed to those people who carry a severe mutation that significantly lowers their IQ. From very large data sets, it has been estimated that as much as 20% of the population carries some such mutation. Some of these people are severely enough affected that they meet the clinical criteria for a diagnosis of "intellectual disability." Others are not that impaired but have much lower intelligence than they would have otherwise (i.e., in the absence of such a mutation).

You can thus model the entire range of IQ as resulting from two overlapping normal distributions—a small one, at the low end, comprising people with severely detrimental mutations, and a larger one, which is the remainder of the population. There is a clear excess of males in the subdistribution at the low end, with twice as many males as females at the lowest point. This is partly due to an excess of those affected by mutations on the X chromosome. Males are more severely affected by X-linked mutations because they have no backup copy of the X chromosome. Mutations in about 200 different genes on the X chromosome can result in intellectual disability, and these account for about 16% of male cases.

This is not quite enough, however, to account for the overall male to female ratio of intellectual disability, which is about 3:2. Another factor that may contribute is the apparently lower robustness of the developing male brain to genetic insults, which also is implicated in increased rates of neurodevelopmental disorders generally. The developing female brain may be better able to buffer the effects of mutations that impact neural development, so that girls with such mutations show less severe effects (see below).

When the small distribution at the low end is removed, the remaining distribution is much more statistically normal (no longer skewed to the left). However, there is still a sex difference apparent at the extreme high end of this distribution. In the top 2% of the distribution (above IQ 130), the ratio of males to females is about 1.4:1. In the top 0.1% (above IQ 140), the ratio is slightly over 2:1. These differences are highly consistent over time and across countries and are seen in children at least as young as 11.

It is not clear what causes this excess of males at the very high end, which occurs without an overall difference in the mean—that is, it's not just that the entire larger subdistribution is shifted to the right in males. One possibility is simply greater developmental variability. The range of phenotypes that emerges from any given male genotype seems to be slightly greater than that for females. This is apparent, for example, in lower facial symmetry of males versus females. This could be due to the effects of testosterone, which add variability, or to slightly lower genetic robustness due to having only one copy of the X chromosome. In brain development, increased variability may lead more males to end up with a more extreme outcome (actually at both the high and low ends). An alternative is that this difference reflects cultural effects, with males with exceptionally high intelligence being recognized earlier and intellectually encouraged more than females. This certainly sounds plausible but it's not clear whether or not it actually holds.

Though mean IQ does not differ between the sexes, there are some differences in specific facets of IQ. For example, males tend to score better on visuospatial tests, such as mental rotation and manipulation of objects. They also score higher, on average, on tests of quantitative reasoning. Females tend to score higher on verbal tests, especially those involving writing. These trends are apparent from childhood and their effects can be seen in standardized test data across 75 countries (from the Programme for International Student Assessment, PISA).[6] These data show a female advantage (higher mean scores) in reading that is fairly large and extremely consistent across countries. They also show a male advantage in mathematics that is smaller and somewhat less consistent. However, these effects are not evenly distributed. The male advantage in mathematics is seen mainly at the high end of the distribution; in the 95th percentile, males outnumber females approximately 2:1 (on average, across countries; data from 2002–12). Conversely, the female advantage in reading is seen mainly at the low end of the distribution (i.e., females and males don't differ that much in reading at the higher levels, but there are many more males at the lower levels).

These differences are modest enough in the main, though reasonably large at the extremes. They have led to a lot of speculation about their possible impact on education and career aptitudes and choices in males and females, in particular the preponderance of males in the physical sciences, technology, engineering, computer science, and mathematics. However, it is extremely difficult to separate their possible influence from the many cultural influences at play, as well as the differences in interests and values between males and females, which may also contribute, perhaps even more so, to career choices.

PSYCHIATRIC DISORDERS

As will be described in chapter 10, the rates of many psychiatric and neurological disorders vary considerably between the sexes. Males show higher rates of autism, ADHD and dyslexia (about a 4:1 ratio), as well as intellectual disability and schizophrenia (about a 3:2 ratio). And they show much higher rates of conditions like stuttering (7:3) and Tourette's

---

[6] Organisation for Economic Co-operation and Development (OECD), *PISA 2015 Results*, vol. 1, *Excellence and Equity in Education* (Paris: OECD, 2016).

syndrome (9:1). Females have a higher prevalence of depression and anxiety disorders, as well as dementia, migraine, and multiple sclerosis (all about a 2:1 ratio). Understanding the reasons for these differences is one of the main motives for studying sex differences in brain development and function.

For the most part, those reasons remain unknown, though we can be confident that they reflect some underlying biological differences, rather than the effects of cultural norms. As mentioned above, there is good evidence that males are less well buffered against the effects of mutations that affect neural development. For example, in female autism patients, the mutations they carry tend to be considerably more severe than those found in male autism patients. This suggests that females are more resilient to the effects of such mutations and consequently it takes a more serious mutation to push them into a pathological state. That may be at least part of the explanation for the greater prevalence of clearly neurodevelopmental disorders in males, but the reasons for the other differences are still mysterious.

That is partly because sex has often been ignored in animal studies aimed at teasing out the underlying biology of these conditions. These often use behavioral abnormalities in animals like rats or mice as proxies for psychiatric symptoms in humans. Such behavioral analyses are typically done using only male animals—at least, that has been the case until recently. The reason is that female rodents' behavior can be affected by their stage of the estrous cycle, which is just another potential confounding factor that most researchers would rather exclude from their experiments. But this practice has meant that sex differences in disease mechanisms and vulnerability have been largely unexplored in animal models.

ROLE OF CULTURE

The preceding discussion presents the compelling evidence for the existence of innate, biologically driven sex differences in brain structure and function, from the macroscopic level visible with MRI down to the biochemical level of gene expression. In this way, humans are like every other mammal—male and female brains are literally made differently.

However, this does not mean that culture plays no role in human sex differences in behavior.

Initial differences due to sex will tend to be amplified by experience, as individuals' own choices affect their experiences. A person's innate interests and aptitudes will tend to lead to selection of activities that suit those tendencies and that foster and deepen them. In this way, the trajectories of boys and girls (or at least some of them) may continue to diverge over time, as they select their own environments and experiences. But they can also be amplified by cultural norms and pressures, in ways that are not driven by individuals' actual traits but by expectations based on group averages.

SOCIAL IMPLICATIONS

In recent years a charge of "neurosexism" has been made against those who claim there are real biological differences in brains and behavior between males and females. This is an understandable reaction to the way such differences have been interpreted by some commentators, as in some way justifying discrimination against women in society. However, you don't need to deny the existence of such differences to argue against sexist interpretations. Sex differences can exist without one form being better than the other.

The existence of group average sex differences also does not justify expectations of individuals. Group average differences are good for predicting things about groups. It is a perfectly reasonable expectation, for example, that most violent crimes should be committed by males, because males are, on average, more violent than females. But group average differences are terrible for predicting things about individuals, especially if the average difference is small, relative to the overall variation in the thing you're trying to predict. It may or may not be true that any given male is more violent than any given female, just as any given male may or may not be taller than any given female.

Similarly, the fact that we can recognize male or female archetypal profiles of psychological traits does not justify their use as stereotypes. Any given male or female will correspond more or less closely to these archetypal profiles—many not closely at all. Prejudging individuals based on the average attributes of groups they belong to is the definition of, well, prejudice.

There seems to be a concern that conceding any ground on the reality of biological differences between the sexes undermines the case for equal treatment. This need not be—indeed it must not be and is not—the case. Equality under the law (in Western democracies at least) does not rest on identity. If it did we would all be in trouble as there is far more variation in the traits we have been discussing *within* sexes than between sexes. The whole point is that moral and legal equality is enshrined in the law—even declared as a "self-evident" truth in the United States—precisely *in spite of such variation.*

While using biological differences to justify sexism is wrong and harmful, ignoring or denying the existence of such differences can also cause harm. This is particularly obvious when it comes to investigating the differences in rates of neuropsychiatric disorders between the sexes. Clearly, something important is going on there, and understanding the general basis for sex differences in brain and behavior between the sexes will be essential in figuring it out.

CHAPTER 10

THE EXCEPTIONS

So far, we've been talking about differences between people that arise from differences in how their brains are wired, in things like personality traits, intelligence, sexuality, and perception. For the most part, those differences are just that—*differences*—contributors to the endless, fascinating diversity of human beings. But sometimes they are more than that. Sometimes differences become disorders.

The term "disorder" implies more than a benign or neutral difference—it entails some degree of suffering or impairment of function based on that difference. Of course, those terms are not independent of societal or cultural effects—the degree of suffering or impairment is not solely down to the individual, but can have as much to do with how society treats or accommodates that person's differences. Autism is a case in point—for people at the less severe end of the spectrum at least, the difficulty in functioning arises in large part because the person's behavior does not fit society's expectations and structures. One could even argue that in some cases, such as conduct disorder or psychopathy, the condition is labeled a disorder because it causes suffering *for other people*.

On the other hand, there are many psychological or neurological conditions that are clearly classifiable as dysfunction, in a more objective sense. Some of these have very selective effects on specific functions, like sleep, circadian rhythms, appetite, speech, reading, or face perception. But others—more common conditions like autism or schizophrenia or bipolar disorder—have much broader consequences, affecting higher-order functions like mood, perception, language, attention, social cognition, even thought itself. The most severe of these strike at the very things that make us human—our reason, our memories, the validity of our perception, our ability to interact with other minds, our sense of self.

The impact of psychiatric and neurological illness is enormous. Being mentally ill makes it more likely that you will be unemployed, live in poverty, be unmarried, and live alone. It greatly decreases number of offspring—for example, people with schizophrenia or autism have only a third as many children, on average, as the population average. And it drastically lowers life expectancy, in the case of schizophrenia by an average of 20 years and for people with autism and intellectual disability by an average of 30 years. Increased mortality rates with psychiatric disorders reflect a very high suicide rate, but also the impact of ongoing mental illness on physical health and lifestyle, including the significant risk of homelessness.

While modern medicine has triumphed over many infectious diseases and made great strides in the molecular diagnosis and treatment of cancer, very little progress has been made in treating psychiatric disorders. Psychotherapy is helpful in many cases but typically more for coping with symptoms than actually curing the condition. And the main drugs used to treat mental illness—antipsychotics, antidepressants, anxiolytics, and mood stabilizers such as lithium—all emerged between the 1940s and 1960s, with almost no new drugs being developed since. In most cases, the existing treatments are only partially effective and can induce serious side effects. These treatments were all discovered serendipitously, and their mechanisms of action remain poorly understood. The explanation for this lack of progress is tragically simple: we have not known the root cause of these disorders—either in general or for individuals.

Genetics is changing that. The incredible advances in genomic technologies over the past decade have revolutionized our approach to and understanding of psychiatric and neurological disorders. Genomic sequencing on an unprecedented scale, across tens of thousands of patients and their relatives, has led to the identification of large numbers of mutations that dramatically increase risk of neuropsychiatric disease. These discoveries are leading to a fundamental change in how we think about these conditions. For one thing, they are clearly not as distinct from each other as we had thought—genetic risk for these disorders is actually highly overlapping. For another, it is clear that each of these categories is hugely heterogeneous—the superficial similarities between patients that get them a diagnosis of, say, schizophrenia hide an underlying diversity of genetic conditions. And, finally, the kinds of genes disrupted clearly implicate defects in neural development as a predominant factor in these conditions.

We will consider below how these discoveries have been made and how they affect the way we conceptualize neuropsychiatric disorders—as predominantly genetic conditions affecting brain development. But first, let's take a moment to consider the long list of other factors that have been proposed as causes of mental illness.

THINGS THAT DON'T CAUSE PSYCHIATRIC DISORDERS

There have been, over the centuries, scores of theories of the causes of mental illness. Possession by demons has been a popular one—especially for episodic symptoms like psychosis and seizures. Unbelievably, it still is—not only are witch doctors and shamans still called on in some parts of the world, but the Catholic Church still has active exorcists. In fact, the idea that schizophrenia and epilepsy are caused by demonic possession is alarmingly widespread among many Western churches. There have even been papers published in the "scholarly" literature lately offering advice on how to distinguish cases of actual schizophrenia from demonic possession.

The field of psychoanalysis exonerated demons and instead took aim squarely at parents—usually mothers—as the cause of psychiatric disturbances later in life. In particular, a perceived cold and detached style of parenting was thought to lead to emotional and social withdrawal of the children and the development of autism and schizophrenia. This idea of "refrigerator mothers" has long since been discredited, but the psychogenic theory lingers on in some quarters, despite the absence of any support for it. We will see below that shared genetic effects more readily explain cases where parental behavior is also affected.

More recently, we've seen all manner of environmental factors blamed for these conditions, especially autism. These include vaccines, genetically modified food, gluten, tablet computers, C-sections, fluoride, air pollution, mercury, pesticides, television viewing, or just not spending enough time outdoors. Part of the reason why people have invoked environmental factors in the case of autism is because it is apparently becoming more prevalent.

For example, rates of autism diagnoses in the United States were 1 in 500 in 1995 but are now over 1 in 100. This rise in diagnoses is indeed alarming, but the key word there is "diagnoses." There is extremely strong evidence that this rise does not reflect a real rise in prevalence but is due instead to greater awareness and better recognition of the condition by parents, teachers, and doctors. Indeed, as rates of autism diagnoses have increased, there has been a *matching decrease* in rates of diagnosis of "mental retardation" or "intellectual disability," suggesting that children who used to receive these labels are now being diagnosed with "autism" instead. This pattern is the exact opposite of what you would expect from a genuine rise in exposure to environmental toxins, which would be expected to increase rates of all neurodevelopmental disorders.

It is worth considering the case of vaccines in more detail, as it illustrates, first, how science works, but second, how misinformation can retain a hold in public consciousness, even in the face of overwhelming contradictory evidence. The idea that vaccines might cause autism first came to prominence with the publication in a highly rated medical journal, the *Lancet*, of a paper describing a small study by Dr. Andrew Wakefield, a British physician, and a number of other researchers.[1] The actual findings of the paper are remarkably unremarkable. Wakefield and colleagues studied 12 children who had been referred to a pediatric gastroenterology unit with abdominal symptoms, along with a cessation of normal behavioral development or even a regressive form of autism. It was claimed that the parents of many of these children noted that symptoms had first appeared around the age the children had received the measles, mumps, and rubella (MMR) vaccine.

On the basis of this highly anecdotal evidence, Wakefield launched a very public campaign claiming that the MMR vaccine was unsafe and specifically that it caused autism. It later emerged that many aspects of this study had been falsified, that it did not have appropriate ethical approval, and that Wakefield had undisclosed conflicts of interest—namely, that he was hoping to profit from new medical tests and litigation-driven testing. This led to the withdrawal of support for

---

[1] A. J. Wakefield, S. H. Murch, A. Anthony, J. Linnell, D. M. Casson, M. Malik, M. Berelowitz, et al., RETRACTED: "Ileal-Lymphoid-Nodular Hyperplasia, Non-specific Colitis, and Pervasive Developmental Disorder in Children," *Lancet* 351 (1998): 637–41. Retraction in *Lancet* 375 (2010): 445.

the conclusions of the paper from Wakefield's coauthors; to the paper being retracted by the *Lancet*, which claims that the journal had been "deceived"; and to Wakefield being struck off the UK medical register.

But the damage was done. Vaccines and autism became linked in people's minds. The horror scenario of a "perfectly normal" child being effectively permanently brain damaged by a vaccine was just too strong and emotive to ignore. And this idea seemed to jibe with the apparent increase in autism diagnoses, which had been dubbed by many as an "epidemic." This has led to a huge decrease in the number of parents vaccinating their children, with the completely predictable consequence that the number of cases of measles, mumps, and rubella has leaped dramatically in recent years. There is a reason we vaccinate against these diseases—they can result in serious long-term consequences, including blindness or deafness, and they have a significant rate of fatality—134,000 people died worldwide from measles in 2015, the vast majority of whom were unvaccinated.

Of course, scientists went to work to see if this supposed link could actually be true. The hypothesis that vaccines cause or increase the risk of autism gives a very clear and testable prediction: children who have received vaccines should show a higher rate of autism than children who have not been vaccinated. That hypothesis has been tested, now many times over, in huge population samples of millions of children. The results could not be more conclusive: there is no increase in the rate of autism among vaccinated children. Not even a tiny one. Not even in children with a family history of autism. Thus, *vaccines do not increase the risk of autism*. This is not a case where "more research" is needed, as is often claimed—the findings are definitive. During the period 2000–2015, measles vaccination prevented an estimated 20.3 million deaths worldwide. There is no evidence to suggest it caused even a single case of autism.

What all of these discussions of hypothetical environmental causes overlook is that *we already know* the main factor underlying these conditions—they are overwhelmingly genetic.

NEUROPSYCHIATRIC DISORDERS ARE GENETIC

It has been recognized for centuries that mental or neurological disorders run in families. Over 2,500 years ago Hippocrates argued that epilepsy was not a "sacred disease," but had a physical cause, and concluded that: "its origin is hereditary, like other diseases." Similarly, ancient Islamic doctors such as Ishaq ibn 'Imran wrote, around AD 900, about hereditary mental illness, and even its overlap with epilepsy. There are scores of examples of famous families afflicted with various forms of mental illness across generations, including, for example, the interrelated royal families of Europe. These are not exceptional cases, as familial aggregation is a strong characteristic of these conditions.

In scientific terms, we can measure how strong that familial effect is by measuring the rate of such conditions in relatives of people who are already diagnosed. For example, the rate of schizophrenia across the population is about 1%, but if you are a sibling of someone with the disorder, your risk of also being diagnosed with it yourself is 10 times that rate. In fact, this increased risk extends across the diagnostic categories of modern psychiatry. If you have a sibling with schizophrenia, your risk of bipolar disorder or autism or epilepsy or many other neuropsychiatric conditions is also substantially increased over the population average. We'll come back to this important point below.

But how can we know that when these conditions do run in families it doesn't just reflect some effect of a shared upbringing or environment? As with so many other traits, twin studies can answer this question. If the familiality were due to a shared upbringing or environment, then MZ or DZ twins would share such effects equally. If one twin is, say, schizophrenic, the rate of schizophrenia in the other twin should not depend on whether the twins are MZ or DZ. But it does. If one of a pair of MZ twins has schizophrenia, the chance that the other will be similarly diagnosed is ~50%. For DZ twins (of the same sex), this rate is only ~15%. For autism, the difference is even more stark—the rate of MZ twins being coaffected is over 80%, while in DZ twins it is only ~20%.

These numbers can be used to estimate the heritability of these conditions—this ranges from over 50% for schizophrenia to over 80% for autism. The exact numbers don't matter that much, partly because such studies don't count cases where one twin has, say, schizophrenia and another has bipolar disorder or epilepsy. They probably therefore underestimate the true risks of illness overall. But there are a few important general conclusions to be drawn from them: first, neuropsychiatric disorders are highly heritable—much more so than conditions like diabetes or heart disease, for example. Most of the variance across the population in who gets these disorders is down to genetic differences. Crucially, this does not mean that such differences are necessarily *inherited* from parents—in many cases they are new mutations generated in sperm or eggs. This can explain why many cases of these conditions are sporadic—still genetic in origin, but not hereditary.

Second, there is no apparent effect *at all* of a shared family environment. This is evident not just from the MZ-DZ twin studies but also from adoption studies, especially of MZ twins who have been reared apart. The rate of schizophrenia concordance in MZ cotwins is exactly the same, regardless of whether they were reared together or apart. And if a person's adoptive sibling has schizophrenia, that person's risk of developing the condition is no greater than any random person in the population. There is thus no support at all for psychogenic theories blaming parenting for these conditions.

And the final conclusion from these twin studies is that *something else*, besides genetics, must be having an important effect. The fact that MZ twin concordance for schizophrenia is about 50% highlights the very important genetic effects. But the other 50%—those MZ twins who are *not* affected—show that a person's genetic makeup does not cause the condition itself, but *risk for* the condition. As for other traits we have looked at, there is thus a *probabilistic* relationship between a person's genotype and eventual phenotype. Whether or not someone actually develops the condition must also depend on other factors. This could include differences in personal experiences or stressors, which may be involved in triggering the acute onset of symptoms in adult-onset disorders. But whether or not a person with a high-risk genotype develops a neuropsychiatric condition likely also depends importantly on the outcome of all the chance events of neural development. More on that later—first let's look at the crucial genetic effects on risk.

THE GENETIC ARCHITECTURE OF NEUROPSYCHIATRIC DISORDERS

Neuropsychiatric disorders have typically been classified into two types: rare forms with known, specific causes, and a great mass of "idiopathic" cases, where a specific cause is unknown. The former include conditions like fragile X syndrome, which is caused by mutations in a specific gene on the X chromosome. This typically results in frank intellectual disability, but the symptoms of the condition can also include autism, attention deficit, or epilepsy, in subsets of patients. It's called a "syndrome" because it also affects other organs, and can be recognized in many patients by a typical facial morphology. Velo-cardio-facial syndrome (now known as 22q11 deletion syndrome) is another example; in this case, about 30% of patients develop psychosis. There are many other conditions like this, with known genetic causes, that manifest with neurological or psychiatric symptoms. Even though the psychiatric symptoms themselves may not differentiate the patients, these kinds of syndromes have been widely considered as quite distinct from the main body of patients diagnosed with idiopathic autism or schizophrenia or epilepsy.

If we take autism as an example, we can see how this works in practice. Typically, if a child is showing the symptoms of autism, he or she may be referred for clinical genetic testing. This will test for a list of rare disorders known to manifest with these symptoms, such as fragile X syndrome, Rett syndrome, Timothy syndrome, tuberous sclerosis, and others. Up until a few years ago, only about 5% of cases of autism could be ascribed to such syndromes, leaving 95% unexplained. There are two ways of thinking about that large group. In the first interpretation, the word "idiopathic" is taken in a limited way to mean exactly what it does mean—we just don't know what the cause or causes are for most patients. This large group could thus be made up of many, many specific genetic conditions like fragile X syndrome that we simply have not yet discovered.

In the second case, that ignorance is treated as a positive finding—our *lack of knowledge* about all those patients is taken to mean that that group makes up a natural kind. In that model, the rare conditions are treated as exceptional and not related to the etiology of "real autism" or "real schizophrenia." This stems from a history of excluding cases that had a known "organic cause," such as syphilis, from the larger category (of schizophrenia, in that case). From a genetic perspective, part of the rationale for making such a separation with the rare genetic disorders is that the inheritance patterns of most idiopathic cases are not so straightforward—the condition may aggregate in families but not segregate in the clear way associated with causation by mutations in a single gene. This suggests that patients in that large group might have a very different underlying mechanism—still genetic, but not so discrete, involving instead the combined effects of multiple genetic variants at once.

We will see below that there is some truth to both these models. The large groups really are umbrella terms for many hundreds of genetic disorders involving specific, identifiable high-risk mutations. More and more of these are being discovered all the time. But the effects of these mutations are strongly modified by additional mutations and variants in each person's genetic background. The genetic complexity of these conditions thus arises both from heterogeneity across patients and from genetic interactions within patients.

FINDING THE CULPRITS

One of the challenges in identifying mutations that confer high risk for neuropsychiatric disease is that they tend to be very rare. This makes complete sense, from an evolutionary perspective. These conditions dramatically decrease both life span and number of offspring, on average, meaning that mutations that cause them should be rapidly selected against. Indeed, the most severe mutations are usually selected against immediately and almost never inherited; they arise almost exclusively as de novo mutations, in the generation of sperm or eggs. Mutations that cause less severe illness, or increase risk to a lower degree, can persist for longer in the population, but still would not be expected to rise to a high frequency.

Rare mutations are hard to find with traditional genetic methods, which have relied on either an obvious and characteristic syndromic presentation to recognize a specific condition, or on very large families with many affected individuals to map a specific gene. But new technologies, applied on a massive scale, are making the identification of rare, high-risk mutations much easier.

The first of these technologies is called "comparative genomic hybridization" and it relies on genomic "microarrays." It is an extremely powerful and cost-effective method to detect small deletions or duplications of segments of chromosomes across large numbers of patients. In this technique, a patient's DNA is collected and labeled with a fluorescent dye. A control person's DNA is also collected and labeled with a different colored dye. Then both sets of DNA are "hybridized" to an array of DNA from the human genome, which is dotted in an orderly way in small segments onto a glass slide. "Hybridization" means that the DNA from a patient's chromosome 1 will stick to the DNA on the slide that comprises chromosome 1, and so on. By comparing the intensity of the two dyes, it is possible to identify small regions of the genome where the patient has either less DNA or more DNA than the control. Usually this means the patient has a deletion of a small segment, so that there is only one copy of that region, or a duplication, so that there are three copies.

Some such copy number variants arise at an appreciable frequency in the population, due to small repeated sequences of DNA, which confuse the machinery that recombines and separates chromosomes during DNA replication. This means it is possible to identify many people carrying the exact same deletion or duplication, so that their effects can be measured statistically, not just in single cases.

It turns out we all carry some background level of CNVs. Indeed, though we may only have a few of these, they actually make up a large part of the genetic differences between people because they encompass so many bases of DNA sequence. That said, most of them don't seem to do anything, mainly because most of them arise in the 97% of the genome that does not encode proteins (i.e., they don't affect genes). But when they do affect genes they can have very deleterious consequences.

When researchers compared autism patients with controls, they found that the patients had a significant excess of CNVs. By looking across many hundreds of patients they could identify particular CNVs that were at much higher frequency in autism patients than in controls, indicating that they dramatically increased risk of the condition. Interestingly, when the same kinds of studies were carried out with patients with schizophrenia or epilepsy or developmental delay or intellectual disability, many of the same CNVs were found. There is now a very long list of these pathogenic CNVs—they include several well-known examples, such as the deletion at 22q11.2 (referring to a specific genomic position on chromosome 22), which is now known to be the cause of what used to be called velo-cardio-facial syndrome, as well as deletions or duplications at 1q21.1, 3q29, 15q11.2 (associated with Angelman and Prader–Willi syndromes), 16p11.2, and many others (see figure 10.1).

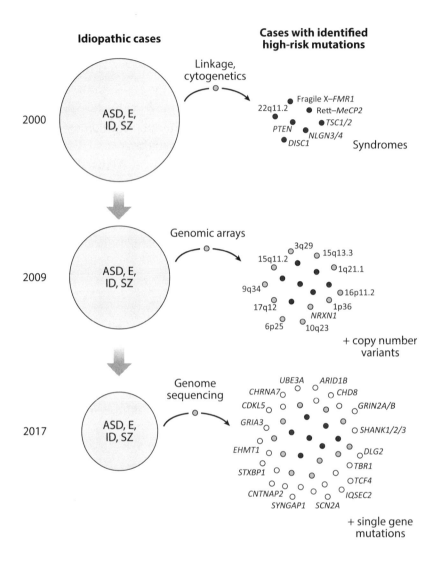

Figure 10.1 Neurodevelopmental disorders. The idiopathic pools of autism (ASD), epilepsy (E), intellectual disability (ID), or schizophrenia (SZ) have been shrinking as more and more specific genetic causes are identified, using new technologies such as comparative genomic hybridization and exome or genome sequencing. (Modified from K. J. Mitchell, "The Genetic Architecture of Neurodevelopmental Disorders," in *The Genetics of Neurodevelopmental Disorders*, ed. K. J. Mitchell [Hoboken, NJ: Wiley Blackwell, 2015].)

Each of these is very rare, usually carried by fewer than 1 in 1,000 people. Collectively, the known CNVs account for 1%–2% of cases of schizophrenia and perhaps 5% of cases of autism. Of course, there may be many other such mutations that we have not yet identified, especially if they are individually rarer or have smaller effects on risk.

The degree of risk associated with each CNV is strongly correlated with whether it tends to arise de novo or is inherited from a parent. This can be determined by also characterizing the DNA of the patient's parents to see if either of them carries the same CNV. Mutations that confer very high risk and cause more severe illness (such as autism and intellectual disability) almost always arise de novo, because affected people tend not to have children, while less severe ones are more often inherited. Determining this fact can obviously have very important consequences for future reproductive decisions in a family.

POINT MUTATIONS

CNVs are just one type of mutation that can cause disease—we happen to know a lot about them because they are easy to detect, thanks to the development of genomic microarrays. More importantly, we can see that *specific* CNV mutations increase risk of neuropsychiatric disorders because they tend to recur over and over again at the same spots in the genome. That means researchers could look across many people with the *exact same mutation* and see that it was greatly enriched in people with disease.

That job gets a lot tougher when we are talking about point mutations— changes to a single letter of the DNA sequence. These occur, at random, all over the genome, every time the DNA is replicated, including when sperm or egg cells are being made. Fortunately, most such copying errors are corrected by a dedicated set of proofreading and DNA repair enzymes, but some creep past that system and become new genetic variants in the population. Up until the past few years, we had no good way of detecting these, unless we had a reason to look in one particular part of the genome (say, from segregation patterns in a large pedigree with many affected individuals). But for most cases of neuropsychiatric disorders we have no such reason—that is, there is every reason to think there is a causal mutation *somewhere* in the three billion letters of the person's genome, we just don't have a reason to look in one place versus another.

That means we need to sequence the whole thing. We need to read the entire code of an individual's genome and then compare that to some reference (or to a large number of other people's genomes) to see where they may have a difference that could be causing disease. This is where the pace of technological change over just the past few years has been truly transformative.

The first human genome to be sequenced—"the" human genome of the Human Genome Project (really assembled from five different people)—took 10 years to complete, with a final draft published in 2003. It involved hundreds of researchers from all over the globe and cost several billion dollars. There were warehouses of sequencing machines working night and day and enormous banks of computers required to process all that data. As I am writing now, in 2017, it is possible to sequence a human genome in a day for under $1,000. Much of this can now be carried out on machines that fit in the palm of your hand and plug directly into your laptop.

This has completely changed genetic research and is poised to change medicine. By sequencing the genomes of thousands of people with intellectual disability, developmental delay, epilepsy, autism, schizophrenia, or related conditions, researchers have been able to detect multiple people with very rare mutations in the same genes. The problem with just looking at the genome sequence of any individual patient is that every person carries a couple of hundred serious mutations that disrupt a gene, altering the protein it encodes or blocking the expression of it altogether. Perhaps only one of these mutations is actually contributing to high risk of the disease, but recognizing which one among all the mutations we all carry is almost impossible, if you only have one patient's sequence. But once you start sequencing hundreds of patients you begin to see repeat hits in the same genes, more than you would expect by chance.

Those efforts have only begun to be scaled up over the past couple of years but are already revealing hundreds of new genetic disorders. Most of these affect genes directly involved in or required for neural development. Each of these conditions is extremely rare, responsible for less than 1% of cases of the general clinical categories listed above.

But collectively they are common—much more common than we ever realized. The ones that have been easiest to spot are, not surprisingly, the ones that cause the severest forms of disease. Many such cases are caused by de novo mutations, for the same reason we discussed in relation to CNVs—people with severe neurodevelopmental disorders tend not to have children, so mutations that cause high risk of such conditions will hardly ever be inherited.

De novo mutations are thus far more likely to cause disease than inherited mutations, which makes it easier to recognize them as the culprits. They can be detected by sequencing a person's DNA along with that of his or her parents. On average, we each have about 70 new mutations that were not present in our parents' genomes. Because these occur at random, and because only ~3% of our DNA actually comprises genes, most of them won't have any effect. The number of de novo mutations that actually hit a gene in any individual is around 1 (it ranges from 0 to 2, compared with about 200 inherited, gene-disrupting mutations). But if you're unlucky, that gene may be one of the several thousand absolutely required (in two working copies) for normal brain development or function.

One of the striking findings from these kinds of sequencing studies is that most de novo mutations happen in the paternal germline (about 75%). There's a good reason for this: in a man's testes, there are stem cells that continue to divide throughout his life—each time they do that there is a small chance of a new mutation happening. Over time, these mutations accumulate in the stem cells and show up in the sperm. By contrast, females are born with all the eggs they will ever produce, so new mutations of this type do not accumulate with age in females (though the chance of abnormalities involving the segregation of whole chromosomes does increase with age). The number of de novo mutations in individuals is therefore linearly related to their fathers' age when they were conceived—offspring born to 40-year-old fathers have about twice as many new mutations as those born to 20-year-old fathers. Not surprisingly, paternal age is also strongly correlated with risk of genetic disease in the offspring. This has long been known for rare genetic conditions of all sorts, but has also recently been recognized for common conditions like autism and schizophrenia, where risk to offspring of fathers over 45 is about four times that of offspring of fathers under 25.

Recent data implicate one other important kind of mutation—ones that happen in the developing embryo itself, known as somatic mutations (because they happen in the body, or soma, rather than in the germline). Mutations that arise in a single cell of the early embryo may be inherited through cell divisions by a significant proportion of the cells of the body, including the brain. If these mutations disrupt development then they may result in a neurodevelopmental disease even if they are "mosaic," or present in only some of a person's cells. Presumed pathogenic mutations of this type have been found in a small percentage of autism patients.

A SPECTRUM OF GENETIC EFFECTS

De novo mutations are the easiest ones to recognize as pathogenic (contributing to disease) because we have fewer of them and they are likely to cause the most severe effects. They explain many of the sporadic cases of disease with no family history. On the other hand, many cases of neurodevelopmental disorders are caused by inherited mutations, which is why they also tend to run in families. These are harder to identify, because they tend to have less drastic and more variable effects, and are far less likely to be acting alone. One way to judge whether a mutation in a person's genome is likely pathogenic is to see whether other people in the population also carry it. Sequencing of tens of thousands of healthy people has provided a map of genetic variation across the population. Like the dog that didn't bark in the night, the real information in that map comes from the mutations we *don't see*.

Many genes show a dramatic absence, or at least a shortage, of damaging mutations when we look across the healthy population. This is not because mutations don't happen in these genes—they happen everywhere in the genome—it's because when they do happen, people get ill or die. These genes are intolerant to genetic variation that knocks out their function. When we see a mutation in a gene like that in someone with disease, it is therefore much more likely that it is pathogenic. And the rarer a mutation is in the general population, the more severe its effects are likely to be.

There is thus a spectrum of genetic variation that can contribute to neurodevelopmental disorders—this ranges from de novo and ultrarare mutations that can have individually large effects, through inherited mutations that have moderate effects and that can persist in populations for some time, to much more common genetic variants that have been around in the population for a long time and that make only tiny individual contributions to risk.

The common ones can be detected using genome-wide association studies. As described in some of the previous chapters, these studies look at the frequency of a given version of a genetic variant in people with a disease, compared with the frequency in people without (controls). If a variant is more frequent in disease cases, it is said to be associated with the disease and is therefore, statistically, a risk factor. This is just like doing epidemiological studies for environmental risk factors. For example, smoking is much more common among people with lung cancer (around 95%) than among people without lung cancer (around 30%). The degree of difference lets you estimate how much of an effect on risk the factor is having. What we measure is how much more likely people are to have exposure to a certain factor, given they have the disease. But this can be flipped around to calculate what is called relative risk—how much more likely they are to have the disease if they are exposed to a specific factor (environmental or genetic), compared with the risk for people who are unexposed. For smoking, the effect size is around 100—people who smoke are around 100 times more likely to develop lung cancer than people who don't.

The challenge in identifying common variants that increase risk of disease is that their effect sizes are usually tiny—on the order of 1.1 or even less. That means people with the common "risk variant" are 1.1 times more likely to develop a disease than people without it. That's literally an almost negligible effect, but not completely. Especially if many common variants combine together—they could then theoretically have a much larger *collective* effect on risk. What it does mean, though, is that we need massive sample sizes to detect that kind of effect with any statistical confidence (i.e., to distinguish a tiny difference in variant frequency between cases and controls as "real," as opposed to just being noise in the data).

That is exactly what has now been achieved in GWAS of schizophrenia—these have recently been carried out on samples of tens of thousands of patients and over a hundred thousand controls. They have identified over 100 spots in the genome where there are genetic variants where one version is at a higher frequency in patient cases than in controls. As with the rare mutations, the implicated genes are highly enriched for genes involved in neural development. As expected, these common variants each have only a tiny effect on risk—most increasing it by less than a factor of 1.1. Collectively, the ones currently identified explain less than 10% of the total variance in liability to schizophrenia, though the total contribution from additional common risk variants yet to be discovered could be much larger.

PUTTING IT ALL TOGETHER

So, what does this all mean? How can we think about all these different types of genetic effects? One way is to think of some proportion of cases being caused by specific rare mutations, while the remainder are caused by the combined effects of many common variants. The idea of the latter model is that we all carry some burden of common risk variants but only when some threshold is reached does this actually cause disease. This dichotomy would entail two really very different things—the former would be quite distinct genetic conditions, while the latter would represent the extreme end of a continuous distribution. But there's really no good reason to separate them out like that—no reason to think there is any real distinction at all between the rare and supposedly common disorders. Quite the opposite in fact—there's every reason to think that multiple genetic variants are at play in each individual, even in cases that inherit a high-risk mutation (see figure 10.2).

First, that would be a typical scenario for any Mendelian disease, that is, one caused by a single mutation. Even for conditions like cystic fibrosis or Huntington's disease, which are always caused by mutations in one specific gene, other genetic variants exist that modify the severity and age of onset and nature of the clinical symptoms. By themselves, these variants don't cause disease—they only have an effect when the person has a rare mutation that causes those conditions. The same is true for all sorts of conditions and we certainly should expect it for neurodevelopmental disorders.

Second, we have direct evidence that specific mutations can manifest in very different ways in different people. Some carriers of specific

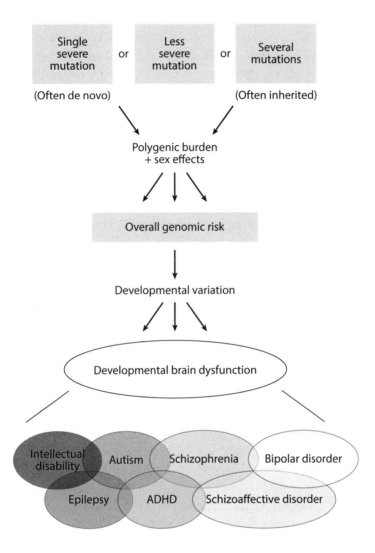

Figure 10.2 The genetic architecture of neurodevelopmental disorders. A single severe mutation (usually de novo) or a number of less severe ones (de novo and/or inherited) can increase risk. Risk is modified by polygenic background and sex, which influence developmental robustness. Randomness in development determines how this risk plays out in any individual. It may result in developmental brain dysfunction, with diverse clinical presentations.

CNVs or specific rare mutations in single genes develop epilepsy, others develop autism, or schizophrenia, and others don't develop any clinical symptoms at all. In some studies, the cases with the more severe presentation have been found to have a second, also rare, mutation somewhere else in their genome, suggesting a combined effect. There are also specific examples for some conditions—notably Hirschsprung's disease, a disorder affecting innervation of the gut—where the effects of a rare mutation are exacerbated by the presence of a specific common variant affecting expression of the other copy of the same gene. The common variant by itself has no real effect—it only increases risk or severity in the presence of a rare mutation. Notably, however, it has a big effect in individuals with such a rare mutation, much bigger than the effect size that would be calculated in a GWAS, as that is averaged across a whole cohort.

Finally, there is a big discrepancy between the risk associated with the rare mutations so far detected and the average rates of illness for MZ twins of persons affected with, for example, autism or schizophrenia. We know from these concordance rates that people who develop, say, autism, *were at really high risk of developing autism*. I know that sounds bizarre, but it's actually a really important point. What you'd want to do to measure someone's overall *genomic risk* of a disorder (given the person in question already has it) is clone that person a bunch of times and see how many of the clones also develop the condition. Say we made 100 clones—maybe only 10 of them would develop it, or maybe 50, or maybe all of them. We obviously can't do that experiment, but we can look at natural clones—MZ twins. We only get one shot for each person, but if we average across many pairs of twins we at least get an average genetic risk associated with the genomes of people who already have the condition. For autism, this risk is about 80%. It's not like those people inherited a low risk and were just unlucky in how things turned out. They inherited an enormously high risk.

By contrast, most of the specific "high-risk" mutations we know of so far only cause autism in a much smaller percentage of carriers—say 30%, if we take fragile X mutations as an example—but this number varies a lot. What that suggests is that the subset of people with those rare mutations *who do develop autism* also have some other genetic factors increasing their overall risk. These could be other rare mutations, specific common genetic variants (as in the example of Hirschsprung's disease, above), or the combined effect of the overall load of lots of such variants.

The last possibility is particularly interesting if we think back to the idea of developmental robustness. The genomic program of development is typically well equipped to buffer the effects of mutations—at least ones that aren't too drastic. But that robustness gets degraded as mutations accumulate. A generally high background level of (relatively mild) mutations, or even common variants, may thus not cause specific disease by itself, but may increase the likelihood that another, serious, mutation will do. One person's genome may be quite able to buffer the effects of a rare mutation, while another person's may be much more vulnerable to its effects.

This relates to the idea that intelligence may be a general fitness indicator. Multiple studies have found that lower IQ is associated with higher risk of developing schizophrenia. This may reflect decreased genomic and neural robustness and a reduced ability to buffer the effects of rare mutations that impair neural development. Higher IQ may, by contrast, be protective. Indeed, several studies have found that aggregate genetic scores from GWAS of cognition tests or of educational attainment (a proxy for intelligence) are correlated with genetic risk of schizophrenia or other forms of mental illness. These findings are consistent with the idea that the background of common genetic variants can influence disease risk at least partly through rather nonspecific effects on system integrity or robustness.

SEX AS A RISK FACTOR

Sex is another major factor that seems to affect a person's ability to buffer potentially pathogenic mutations. Many neurodevelopmental disorders are more common in males than females. The male to female sex ratio in autism is about 4:1, and the same is true for ADHD and dyslexia, while for schizophrenia and severe learning disabilities it is 3:2. (By contrast, depression and anxiety disorders are more common in females.) Part of the male excess is due to cases caused by mutations on the X chromosome. Because males only have one copy of the X, a mutation in a given gene on the X can have serious consequences, as there is no backup copy. But X-linked cases are not sufficient to explain the overall male excess.

Being female seems to offer some more general protection against the effects of rare mutations (or being male increases vulnerability). For example, when groups of patients with autism were analyzed for potentially pathogenic CNVs, it was found that the female patients had much larger CNVs, which disrupted many more genes. This suggests that it takes a more severe mutation to cause autism in females than in males. On similar lines, when a pathogenic CNV was found to have been inherited in patients with autism, it was much more likely to have been inherited from the mother than the father. Again, this suggests that females are better able to tolerate such mutations and more likely to pass them on, while males are more likely to be severely affected and thus less likely to have children.

The reasons behind this effect are not known. It could reflect increased robustness of the genetic program, due to the extra copy of the X chromosome in females, which would buffer effects of all the genetic variants on that chromosome. But if that were true it might be expected to be more evenly manifested across all developmental disorders, while what we actually see is quite a range of sex ratios for different conditions. An alternative is that the male brain is made more vulnerable by the effects of testosterone or by the direct influence of genes on the X or Y chromosomes on brain development (see chapter 9). This might have quite uneven effects on different neural systems, explaining the divergent sex ratios across different conditions. If this is true, then it is likely the presence of the Y chromosome, not the absence of an X, that explains the effect. Differences in either the wiring of or the hormonal milieu in the male brain may make it more susceptible to neurodevelopmental insults. Indeed, this increased vulnerability in males is seen even in conditions that are sometimes caused by obstetric complications, such as cerebral palsy.

THE TYPES OF GENES RESPONSIBLE

The genes so far implicated—either by rare mutations or common variants—are, as a class, strongly enriched for genes expressed in the fetal brain. Many of them have direct roles in neural development—for example, in regulating the expression of other genes in various cell types, controlling the migration of neurons and orchestrating the detailed cellular architecture of various brain regions, directing the guidance of growing nerve fibers or the specification of synaptic connections, or mediating the biochemical and cellular processes of synaptic plasticity and activity-dependent refinement.

Here are a few examples: *NRXN1* (neurexin-1) encodes a protein involved in specifying synaptic connectivity—in fact it encodes multiple different versions, which are expressed in highly specific ways in different cell types. These various forms of the NRXN1 protein sit on the surface of nerve cells and act as a receptor for partner proteins expressed on other cells—when a match is detected, a synaptic connection will be stabilized between them. Deletion of even one copy of the *NRXN1* gene can impair the formation of neural circuits in specific ways that are still being worked out and that somehow result in a range of clinical conditions, including autism and schizophrenia.

Mutations in the *CHD8* gene are a rare cause of autism and other neurodevelopmental conditions, often associated with macrocephaly (a large brain and head). This gene encodes a protein that helps regulate the expression of thousands of other genes during neural development, including genes associated with variation in brain volume. However, exactly how mutations in *CHD8* affect all these other genes and how that leads to these clinical outcomes will clearly be challenging to figure out.

The *SHANK3* gene encodes a protein that acts as a scaffold to help organize the complex molecular machinery at synapses. Deletions of this gene cause Phelan-McDermid syndrome, which is characterized by developmental delay and intellectual disability, but *SHANK3* mutations have also been found in patients with nonsyndromic autism and schizophrenia. Mutations in *SHANK3* alter the distribution of other proteins, especially ion channels, at synapses, changing the electrical properties and patterns of neurotransmission. These relatively subtle changes in neural circuits during development can ultimately lead to quite profound disruption in the function of various brain systems; though, as with the other genes referred to, how that happens is currently still a mystery.

Other implicated genes have less direct functions in neural development but are still required for it to proceed normally. For example, many forms of intellectual disability and even some cases of autism and psychosis can be caused by mutations in metabolic enzymes. These proteins carry out the many, many steps of cellular metabolism—converting chemicals from one form to another in complicated biochemical pathways. They are thus not "neurodevelopmental" genes, but if their functions are impaired the indirect effects on brain development can be profound.

You can see just from those few examples that the functions of these genes are quite diverse, and there are hundreds of others that I could have chosen. There are, simply put, many ways that brain development can go wrong. These discoveries highlight several important facts. First, they show directly that the broad diagnostic categories used in neurology and psychiatry really are umbrella terms for hundreds, or perhaps thousands, of distinct genetic conditions. That doesn't mean the genetics in each case is simple—far from it, as we will see below—what it does mean is that the genetic causes are *extremely heterogeneous*. The clinical labels used—intellectual disability, autism, schizophrenia, epilepsy—thus do not represent unitary conditions at all. Second, these categories are also not distinct from each other, at least not in terms of genetic causes.

ON THE VALIDITY OF PSYCHIATRIC CATEGORIES

One of the most striking revelations from recent genetic discoveries is that the clinical effects of high-risk mutations do not respect the boundaries of psychiatric diagnostic categories. Without exception, all of them increase risk across a range of disorders, including autism, ADHD, epilepsy, intellectual disability, schizophrenia, and bipolar disorder. This fits with the overlapping risk for these disorders observed between relatives in epidemiological studies.

A lot of effort in psychiatry has gone into defining, and repeatedly redefining, these categories, based on clusters of clinical symptoms. They are taken as reasonably valid divisions with both descriptive and predictive value, for things like medication responsiveness, for example. However, even at the clinical level, the various neuropsychiatric diagnostic categories have significant overlap in symptoms and commonly occur together in individual patients. Many patients also drift from one diagnosis to another over time. There is thus some question as to how well these categories define specific disordered states. It is probably best to think of them as "open constructs"—not tightly delimited by a rigid set of criteria, but looser definitions of "types of things" that may be recognized by general similarity. The validity of the distinctions based on end states is thus somewhat questionable, though the diagnostic categories retain some clinical usefulness.

But viewed from the other end—focusing on their origins, rather than the outcomes—it is now abundantly clear that these conditions have a common or at least highly overlapping etiology. In terms of what causes them, they are not so distinct at all. There is thus no such thing as "the genetics of autism" or "the genetics of schizophrenia"—these are different aspects of the same thing. Given the evidence that these conditions have their origins in disturbed neural development, an alternate, encompassing term is the genetics of "developmental brain dysfunction."

This highlights the causal origins but allows that they may manifest in diverse ways in different patients. Indeed, it is possible to arrange the various outcomes on a continuous spectrum of overlapping clinical syndromes, rather than as discrete categories with no relation to each other. This spectrum runs from intellectual disability and severe developmental delay at one end, through autism (which itself has a broad spectrum of severity), to later-onset conditions such as schizophrenia, schizoaffective disorder, and bipolar disorder.

WHY DO NEUROPSYCHIATRIC DISORDERS PERSIST?

Many people have speculated as to why conditions like autism, schizophrenia, or bipolar disorder persist in the population. Their continued presence seems to demand an explanation, in particular an evolutionary one. A number of theories have been proposed for why these conditions—or the genetic variants predisposing to them—may actually be advantageous in some ways, which could counterbalance the effects of negative selection against disease-causing variants. Maybe in ancient societies there was something beneficial about these conditions, or at least having a high load of risk variants. For example, maybe the same variants that lead to schizophrenia or bipolar disorder sometimes alternatively lead to creative genius, of the kind seen in great poets or artists. Maybe the variants that cause autism can lead to mathematical talents or other kinds of intellectual genius.

Well, maybe, but natural selection doesn't care how creative you are, unless what you are creating is children (who survive to also have children). And relatives of people with these kinds of conditions do not tend to have more children than average, which would have to be the case to offset the very large decrease in number of children of the affected people themselves. There is thus no evidence for this kind of balancing selection actually happening, either now or in the past.

But, more fundamentally, these kinds of theories are offering solutions to a problem that doesn't exist. The inference is that, because these conditions persist in the population at a fairly steady rate (say around 1% for both schizophrenia and autism), the genetic variants that cause them *must also persist*. This would indeed require an explanation, but it doesn't happen. Individual mutations conferring high risk for these conditions are very rapidly and efficiently removed from the population by negative selection. The conditions persist because new ones keep getting generated. There is thus a balance between mutation and selection that maintains an equilibrium rate of prevalence.

And that prevalence level is determined by the "mutational target"— how many genes there are in which a mutation will result in that condition. For these kinds of neurodevelopmental conditions, that number may be well over 1,000. Now *that* requires an explanation. Why is the brain so delicate? Why couldn't evolution craft a more robust genetic program of development? Well, this is speculative too, but it seems likely that the runaway process that led to our increased brain size, complexity, and intelligence carried a price of increased vulnerability to mutations. Like any piece of machinery, the more sophisticated it gets, the more ways there are for it to break down.

Evolution can't future-proof things. It can build in robustness to noise at the molecular and cellular levels—indeed, it must—and this can indirectly confer resilience to genetic variants with small molecular effects. But it can't anticipate all the serious mutations that might happen in the future. If the benefits from an upgrade in complexity are great enough, it will be strongly selected for, even if that comes at the expense of a small proportion of future individuals. We are, in effect, early adopters of a new operating system, but the downside is we are constantly beta testing it.

However, this only explains part of what needs explaining. The fact that it takes a lot of genes to build a human brain can explain why neuro-developmental disorders are collectively common and persist at that level. But it doesn't explain why they manifest in such strange ways. If they just caused a decrement in function—like intellectual disability—it would be perfectly understandable. But they don't—the symptoms of schizophrenia or autism or bipolar disorder or epilepsy are not simply decrements in normal functions. They are qualitatively novel states.

EMERGENCE OF SPECIFIC PATHOLOGICAL STATES

Why do the very particular types of strange behaviors in autism—narrow, intense interests; insistence on sameness; stereotyped repetitive motor behaviors; poor social functioning—arise and not others? How do neural circuits end up in an out-of-control state of resonating hyperexcitability in a seizure? Why, in psychosis, do people see and hear things that don't exist; why do they form false beliefs, with very definite trends in their content? This is not simply circuits or systems not performing well—they are actively misbehaving, in a fairly narrow set of ways.

So, we must ask ourselves, why do those symptoms emerge instead of all the other ways we could imagine things going wrong? There is a remarkable convergence onto a fairly small number of recognizable symptom clusters; that's why we have diagnostic categories in the first place. How is it that mutations in so many different genes converge onto *these specific outcomes*? This bears on the related question of what are genes *for*?

Geneticists often use the shorthand phrase of *a gene for* some genetic disorder—it could be deafness or dwarfism or cancer or autism. What this phrase actually means is that *mutations in that gene* can cause that condition. Unfortunately, it sounds like it means something else—like the function or purpose of the normal version of the gene is to cause the condition. Of course that's not the case at all; if anything, the normal function of the gene is often the opposite of what happens when it is mutated. Mutations in a gene involved in promoting bone growth may lead to dwarfism. Mutations in genes involved in regulating cell division may lead to cancer.

However, it is not always the case that the effects of mutation in a gene relate so directly to its normal function. Indeed, if the effects are measured at a level far removed from the molecular functions of the encoded protein, then there may be no correspondence at all between them. This is likely to be especially true for psychiatric disorders, where the symptoms are defined at the highest levels of mental function—perception, mood, memory, language, thought.

The genes implicated in conditions like autism and schizophrenia are not genes for social cognition or for regulating anxiety, they are not genes for seeing only things that actually exist or for maintaining a co-herent stream of thought; they are, for the most part, simply genes for building a brain. The psychological effects that arise when these genes are mutated are *emergent*, not a direct result of the absence of that gene's function. In many cases, the symptoms arise not because the gene is not currently functioning, but because *it was not* functioning, while the brain was developing.

A given mutation may have effects on quite specific cellular processes, such as the migration of specific types of neurons, or the formation of specific synaptic connections, or the regulation of synaptic plasticity in developing circuits. But, due to the contingent and self-organizing nature of brain development, these primary defects will have cascad-ing effects over subsequent processes. If the initial connections are not formed properly, then the patterns of activity that drive the elaboration of neural circuits will necessarily be altered. Any disorganization at early stages may thus propagate through activity-dependent processes, alter-ing the circuitry of interconnected areas throughout the brain. This may lead to the ultimate emergence of pathological states, even qualitatively novel ones (see figure 10.3).

To take a well-studied example, manipulations that affect the de-velopment of the hippocampus, if performed very early in the life of a rat—within a week after birth—can lead this structure to become hyperactive. As a result, it drives activity more strongly in the develop-ing midbrain, in a region that releases dopamine. This leads, in turn, to changes in the striatum and prefrontal cortex—targets of these do-pamine neurons—and to the emergence of a state that mirrors aspects

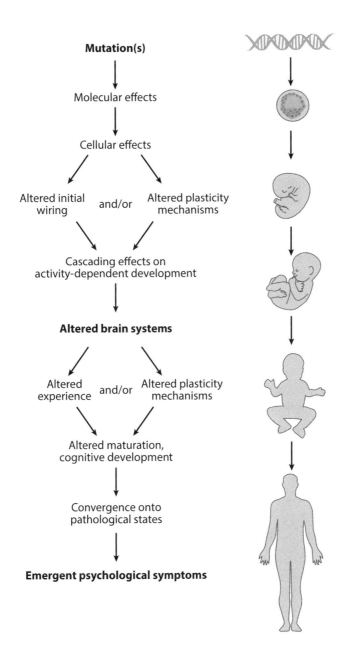

Figure 10.3 The emergence of psychological symptoms. Direct effects of genetic variants at the cellular level can impair subsequent development, with cascading effects over time, sometimes leading to the eventual emergence of psychological symptoms.

of psychosis in humans (which is characterized by alterations in do-pamine signaling, among other things). Importantly, manipulations to the hippocampus later in life, after these circuits have wired up to-gether, do not produce this effect—it is an emergent property of the *developing* brain.

In the face of some insults, the self-organizing properties that nor-mally channel neural systems into a typical outcome may instead chan-nel the brain into some alternate stable state. That there is only a limited set of such stable states—or failure modes—may not be that surpris-ing, if we consider the many nonlinear interactions, contingencies, and intercoupled feedback loops that characterize the developing brain.

This brings us back to the question of the nongenetic sources of vari-ance in who develops these kinds of conditions. They are very highly heritable (i.e., most of the variance is due to genetic differences), but not completely. The fact that some of the variance is not genetic has been taken to mean it must be environmental in origin, though the search for environmental risk factors has not been very fruitful. An alternative is that while individuals inherit a certain probability to develop a condi-tion, whether or not they actually do depends also on chance events during development, which may be amplified by the self-reinforcing processes of neural development to channel an individual phenotype down one trajectory or another.

These cascading effects may continue over the course of later cogni-tive development, by altering the nature of a person's experience—both objectively and subjectively—in ways that then amplify initial differ-ences or further channel an individual's development along a specific trajectory. For example, a child with autism may start out life with less innate interest in other people's eyes. The child may thus miss out on the social cues of shared gaze that are so important to language develop-ment and communication. This may lead to deficits in social cognition or delays in language acquisition, even though language systems were not directly affected by the causal mutation.

Though the details have yet to be worked out, the answer to why the observed phenotypes in neurodevelopmental disorders converge onto a limited set of particular pathological states is thus more likely to lie in the properties of the developing brain than in the molecular or cellular functions of the implicated genes.

CLINICAL IMPLICATIONS

Whole-genome sequencing of patients with neurodevelopmental disorders has the potential to define hundreds of new, rare genetic conditions and to provide genetic diagnoses for many more patients than currently possible. This can provide clarity over the prognosis and expected clinical course, and bring a welcome end to what is often a long and frustrating diagnostic odyssey. A definitive genetic diagnosis can also inform symptom management or surveillance for known complications and allow condition-specific family support.

It should be emphasized, however, that the genetics of these conditions is usually complex, often involving more than just a single mutation. The cases caused by very high-risk mutations have been the easiest to recognize and define, but even those mutations are affected by modifiers in the genetic background, and have quite variable outcomes. The challenge for the future is to define how multiple mutations interact in individuals to generate overall risk. This will mean moving from analyses of very rare mutations—already challenging—to assessing the risk of ultrarare or even unique combinations of genetic variants.

That said, it will be possible in many cases to identify a main culprit—a high-risk mutation without which the individual would most likely not have the clinical condition. For very rare disorders, such a diagnosis importantly also allows the identification of multiple patients around the world suffering from the same condition. This is of huge benefit in defining the clinical characteristics of the condition and also allows patients and their families to contact each other and share their experiences. This kind of contact has resulted in the growth of support and advocacy groups for a growing number of rare disorders, which can play a strong role in driving research forward.

In addition, it allows estimation of the genetic risk to relatives, including, in cases where one child in a family is affected, the risk to additional offspring. If a mutation is found to have arisen de novo, then there is no additional risk to future offspring. If it was inherited, then risk may be as high as 50%. Knowing the causal mutation does, however, provide the means for genetic screening, either prenatally or prior to implantation of embryos generated by in vitro fertilization. Genetic information is already routinely used to screen for many inherited conditions or to detect recurrent de novo conditions such as Down syndrome. These applications are likely to become more common as more conditions are defined. We will look in chapter 11 at the ethical and moral implications of these genetic technologies.

A genetic diagnosis may also allow much more personalized medical treatment. For example, mutations affecting metabolic enzymes are an important subset of neurodevelopmental disorders. In some cases, such as phenylketonuria, the symptoms may be ameliorated or prevented by careful dietary restrictions. For other conditions, such as epilepsy, a genetic diagnosis can indicate or contraindicate specific medications. For example, Dravet syndrome is a rare cause of severe infant epilepsy. It is usually caused by mutations in the gene *SCN1A*, which encodes a sodium channel protein. It is now known that the most commonly prescribed anticonvulsants, which block sodium channels, actually make seizures worse in these patients and should thus be avoided.

For most conditions, however, the discovery of the causal mutation is just the first step in a very long road to developing treatments. There may be some cases where a specific biochemical pathway affected by the mutation can be directly targeted by a drug. If the symptoms of the condition arise due to an ongoing deficit in that pathway, then such a treatment may have some benefit. This could be true, for example, for metabolic conditions; for mutations affecting ion channels, which acutely disrupt the balance between excitation and inhibition in the brain; or for mutations affecting ongoing processes of synaptic plasticity, as in fragile X syndrome. The development and clinical testing of such drugs is typically a very long process, however.

For many conditions, the highly indirect relationship between gene function and emergent neuropsychiatric symptoms will make such direct therapeutic options impossible, especially where the mutation had its primary effects during early brain development. However, it may be possible even for these conditions to define the emergent state at the neurobiological level—for example, by using animal models that recapitulate the causal mutation. If we can achieve sufficient understanding of these states then we might be able to devise treatments to correct or compensate for specific circuit abnormalities. These could include drugs that target specific channels or receptors, but may also include electrical interventions such as deep-brain stimulation (already being used for Parkinson's disease and obsessive-compulsive disorder) or even customized behavioral therapies, possibly incorporating neuro-feedback technologies.

Finally, precision genome-editing tools (such as the CRISPR/Cas9 system) may eventually offer the means to actually correct the genetic defect in embryos. This technique provides the means to precisely change the DNA sequence at a specific position in a living cell. This is already being trialed in blood cells as a therapy for severe combined immunodeficiency and related conditions. However, performing it in human embryos still faces serious technical hurdles and, moreover, raises profound ethical and moral issues, as any such changes would also be inherited by that person's future offspring. For the moment, genetic screening is a much more real possibility, but genome editing will likely become a real possibility in the near future—how we decide to deal with that option remains to be seen.

CHAPTER 11

IMPLICATIONS

Well, we've reached the end. Time to look back at what we've discussed and draw a few final general conclusions. The first section of the book presented a broad picture of how genetic and developmental variation together cause innate differences in psychological traits. The second section considered these issues in relation to specific areas, exploring the diversity of human faculties affected and what is known of the underlying mechanisms in each case. We are only beginning to unravel these details but we know enough to sketch out broad conceptual frameworks for how genes affect these diverse traits. Hopefully the general principles described will stand up to the test of time and will prove useful in interpreting future discoveries.

I will reiterate and expand on some of these general principles below and, especially, emphasize the complexities and subtleties in the relationship between genetic variation and variation in psychological traits. I will also try to highlight not just what the scientific findings mean but also what they don't mean, to clarify or preempt any simplifications, misunderstandings, or overextrapolation.

And, finally, I will consider some important implications of these findings across a range of societal, ethical, and philosophical issues. The genetic and neuroscientific discoveries described in this book are poised to change our ability to control our own biology, as well as our view of our selves and of the nature of humanity. We would do well to consider the potential ramifications now, because the pace of discovery will only accelerate.

WHAT GENES ARE FOR

Twin, family, and population studies have all conclusively shown that psychological traits are at least partly, and sometimes largely, heritable—that is, a sizable portion of the variation that we see in these traits across the population is attributable to genetic variation. However, as we have seen in the preceding chapters, the relationship between genes and traits is far from simple.

The fact that a given trait is heritable seems to suggest that there must be *genes for that trait*. But phrasing it in that way is a serious conceptual trap. It implies that genes exist that are dedicated to that function—that there are genes *for* intelligence or sociability or visual perception. But this risks confusing the two meanings of the word gene: one, from the study of heredity, refers to genetic variants that affect a trait; the other, from molecular biology, refers to the stretches of DNA that encode proteins with various biochemical or cellular functions.

If the trait in question is defined at the cellular level, then those two meanings may converge—for example, differences in eye color arise from mutations in genes that encode enzymes that make pigment in the cells of the iris. They really are genes *for eye color*—that is the job of those proteins. Similarly, mutations that cause cancer, where cells proliferate out of control, mostly affect genes encoding proteins that directly control cellular proliferation. That kind of direct relationship between the effects of genetic variation and the functions of the encoded gene products makes complete sense if you are looking at effects on a cellular level. But it makes no sense if you are talking about the emergent functions of complex multicellular systems, especially the human brain.

These emergent functions rely on the interactions of hundreds of different cell types, organized into highly specified circuits, first at the local level of microcircuits and then at higher and higher levels of connectivity across brain regions and distributed systems. It requires the actions of thousands of genes to build these circuits and mediate the biochemical functions of all the component cells. Variation in any of those genes could, in principle, affect how any given neural system works and manifest as variation in a behavioral trait.

The fact that a trait is heritable means only that there are *genetic variants that affect that trait*. But for the kinds of traits we are talking about, most of those genetic effects will be highly indirect. Natural selection may see such variants as "genes for intelligence" or "genes for sociability" because natural selection only gets to see the final phenotype. That does not mean that the encoded gene products are directly involved in that psychological function. There are no genes for complex psychological functions—there are neural systems for such functions and genes that build them.

This has important consequences for understanding the relationship between genotypes and psychological phenotypes. First, a lot of the variation in mature function stems from differences in how the neural systems develop. Our brains really do come wired differently—literally, not metaphorically. Here, the effects of genetic variation combine with those due to inherent noise in the cellular processes of development themselves. The program encoded in the genome can only specify developmental rules, not precise outcomes. And the more genetic variants there are affecting that program, the greater the variability in outcome will be. Any given genotype encodes a range of potential outcomes but only one—a completely unique individual—will actually be realized.

Second, the genetic architecture of such traits is not as modular as often thought—any given neural system can be affected by variation in probably hundreds of genes. Conversely, variation in any given gene will typically affect multiple functions. In fact, even the neural systems are not as modular and dedicated as once believed—most cells, circuits, or brain regions can flexibly engage in various tasks by communicating with different subsets of other cells, circuits, or regions. When we open the lid of the black box and look inside, we should not expect to see lots of smaller black boxes. It's a mess in there (see figure 11.1).

And, finally, the genetic variants that contribute to any given trait are highly dynamic over time. Natural selection has spent millions of years crafting the finely honed machine that is the human brain, and it's not about to stand back and let it all go to pot. New mutations arise all the time, but those that impair evolutionary fitness—by affecting survival and reproduction—are selected against, with the ones with most severe

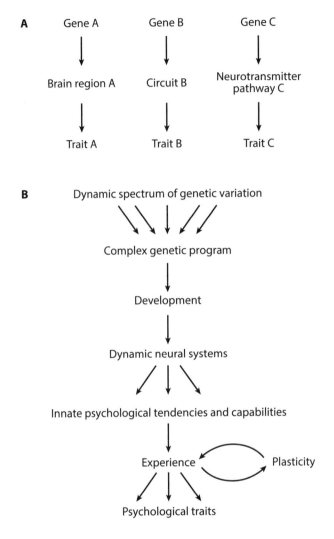

Figure 11.1 Simple versus complex traits. **A**. An overly simplistic view of the relationship between genes and behavioral traits, mediated by direct effects on particular brain regions, circuits, or neurotransmitter pathways. **B**. A more realistic view of the complex genetic architecture of behavioral traits.

effects rapidly disappearing from the population. This means that most traits will be dominated by rare mutations that wink in and out of existence in populations over time, rather than a pool of standing variation that just gets reshuffled from generation to generation. Moreover, the effects of many such variants (rare and common) will interact in complex ways in any given individual. All of these factors have important implications for the possible application of genetic information in predicting the traits of individuals.

GENETIC PREDICTION AND SELECTION—THE NEW EUGENICS?

The complexities described above will make it more challenging to identify specific genetic variants associated with specific psychological traits. And, even where they are identified, predictions of phenotypes based on genetic information will remain imperfect. The effects of single mutations almost always vary across individuals, depending on other genetic variants in their genomes, and multiple variants will often interact in complex ways. It may be possible to derive an average risk of a condition or an average value of a trait from population studies, but it will be very difficult to predict accurately in any individual, who will inevitably have a previously unseen combination of genetic variants in their genome. Moreover, developmental variability places a strong limit on how accurate genetic predictions can ever be, as it means that genotype-phenotype relationships are not just limited by current knowledge but are *essentially probabilistic* and will therefore never be predictable with complete accuracy.

However, genetic information doesn't have to be 100% accurate in predicting traits or disorders for it to be useful. Even mutations that merely increase the risk of a condition, or variants that tend to increase or decrease the value of a trait, will likely be deemed actionable and may be used in reproductive decisions and possibly in other areas. We already know, for example, of hundreds of genes that, when mutated, increase the risk of neurodevelopmental disorders, manifesting as intellectual disability, autism, epilepsy, schizophrenia, or other diagnostic categories. Many of these mutations also affect intelligence more generally, even in people not severely enough affected to be clinically diagnosed, and other genetic variants with subtle effects on intelligence are also being discovered. A number of mutations have been associated with impulsivity, aggression, and antisocial behavior—ones causing other personality disorders, such as being a psychopath, are sure to follow. And it is only a matter of time before mutations affecting other traits, like sexual orientation, or conditions like synesthesia or face blindness are identified.

With this knowledge will come the opportunity to act on it. The most obvious way in which genetic information will be used—indeed, the way in which it is already being used—is in prenatal screening of fetuses or preimplantation screening of embryos generated by in vitro fertilization (IVF). Genetic screening of fetuses for chromosomal conditions such as Down syndrome is routinely done in many countries, and this could readily be extended to screen for deletions or duplications associated with neurodevelopmental disorders more broadly. It is even now possible to sequence the entire fetal genome noninvasively, by sampling the small number of fetal cells that circulate in the maternal bloodstream. This will allow the identification of potentially disease-causing single base changes to the DNA sequence, not just large chromosomal aberrations. The expected consequence, where such measures are available, is a concomitant increase in the number of terminations and a decrease in the number of children born with these conditions.

IVF provides even greater scope for the use of genetic information, as multiple embryos are generated at once. It is quite routine to perform genetic testing on embryos to screen for chromosomal anomalies, especially in older parents or ones with a history of miscarriages. And genetic testing is also done in cases where one or both parents are carriers of a known specific mutation associated with a disease. In such cases, unaffected embryos can be chosen for implantation. Genetic testing is also used in some jurisdictions to select embryos by sex, to screen for immunological compatibility with a previously born child in need of an organ transplant (so-called "savior siblings") or even, in cases where both parents have a condition like deafness or dwarfism, to select *for* the presence of mutations that result in that condition in their children.

As with fetal screening, the range of genetic variants and number of associated conditions or traits that can be screened for will only increase with time. Currently, a limiting factor on how many things can be screened out is the number of eggs that can be obtained for fertilization.

This may change with the recent development of techniques to generate large numbers of eggs in the lab from cultured stem cells (themselves derived from a person's skin cells, for example). This kind of approach is costly, but could mean that hundreds of embryos could technically be generated and screened at once, changing the possible scope and pace of genetic selection.

Clearly, the ethics of the use of genetic information in this way merits some consideration. This is especially true given the dark history of eugenics and its association with the science of genetics. Francis Galton, whom we met in earlier chapters, coined the term "eugenics" in 1883 to refer to the idea of selective breeding in humans to "improve" the genetic stock of the population. He argued that what had been done in dog breeding, with a rapid response to strong selection, could just as well be done in humans. In particular, he bemoaned what he saw as the reproductive excesses of the lower classes in Great Britain that threatened to flood the gene pool with inferior genetic variants, which would, over time, degrade the average capabilities of the population. To counter this threat, he advocated programs to encourage people of higher intellectual ability to breed early and often.

In the early 1900s eugenics achieved wide popularity in Britain and especially in the United States. Prominent geneticists like Charles Davenport, and even celebrities like the aviator Charles Lindbergh, threw their weight behind it and it came to be entangled with issues of race and immigration. Davenport established the American Breeders' Association, with the rather chilling mission to "investigate and report on heredity in the human race, and emphasize the value of superior blood and the menace to society of inferior blood."[1] Rather than just promoting breeding of those with supposedly high quality genes, American eugenic policies focused on preventing breeding by those with qualities deemed inferior. This included marriage bans and forced sterilization of the "feebleminded" and even people with epilepsy to prevent the passing on of the "genetic taint," to use the terminology of the day. Such policies persisted as late as the 1970s in some US states. The underlying principles of eugenics and the idea of racial superiority were warmly embraced in Nazi Germany and used to justify many of the horrors that followed.

---

[1] F. R. Marshall, "The Relation of Biology to Agriculture," *Pop. Sci.* 78 (1911): 553.

Eventually, the principles of eugenics and the policies of socially engineering the gene pool, from encouraging marriage to outright genocide, were rejected by modern societies. There are some schemes in place in certain countries or ethnic groups where specific genetic conditions are especially common that encourage or require people who wish to marry to undergo genetic testing for those specific mutations. But the kind of broad, government-imposed restriction of breeding opportunities based on undesirable traits seen at the height of the eugenics movement is no longer in place in any country.

However, in its place is emerging a different idea, one based instead on the principles of personal or parental choice. This is seen by many as a natural extension of already existing options for reproductive choices available in many countries. The argument goes that if termination of pregnancies or selection of embryos for implantation is permissible at all, there is no reason that such choices could not be made on the basis of genetic information. Different states have taken different views of this. For example, preimplantation testing for genetic conditions is limited in the United Kingdom to a specified list, though this continues to grow over time. And testing for sex is permissible in the United States, but not in most European countries.

There are no easy answers here. You could argue that if no one is harmed (and embryos not being implanted don't count, because that happens all the time anyway), then use of any genetic information should be permitted. On the other hand, this touches on much wider issues. Choosing based on clear medical grounds is one thing; choosing between two healthy embryos is another. Do parents really have the right to choose the traits of their child? Does this change the nature of the relationship? Does it incur some responsibility on parents for the traits of their offspring, if they either have or have not selected them? Will it alter how people who are born with conditions that are otherwise typically screened out are perceived and treated in society? Will changing practices put pressure on parents to make certain decisions?

I am not taking or advocating any position here—all of this is just to highlight the fact that these ethical issues exist and merit some discussion. And as the pace of genetic discoveries advances and new technologies develop, new issues will arise—ones that we may not even have conceived of yet. For example, the recent development of highly precise genome editing technologies (the CRISPR/Cas9 system, referred to in chapter 10) opens the possibility to go beyond screening and begin genetic modification of human embryos. That is currently outlawed, where it would lead to the modification being passed on through the germline, but this could change. Societies will have to grapple with these issues, and make principled decisions as to what should be permitted. We would do well to consider the implications before they happen or we will be closing the barn door after the horse has bolted.

One particularly touchy issue is the idea of selecting for intelligence. We already select against mutations that cause intellectual disability. It seems a small step to extend this to allow selection for intelligence across the typical range, if we have the means to do it. Indeed, some would argue that there is nothing to discuss, that it is obvious that we should allow parents to make that choice if they wish. This can swing back into eugenics territory very quickly, though. You can argue that being more intelligent will be better, for the person involved, than being less intelligent, all other things being equal. After all, higher intelligence is associated with greater general health, better life outcomes across a range of measures, and increased longevity. But that does not mean that intelligent persons (or embryos that will become more intelligent persons) *are better than* or of "higher quality" than less intelligent persons, as some commentators have asserted. Nor does it mean that it would be better *for society* if average intelligence were increased. That's right back to the driving principles of Galton and Davenport.

From a technical point of view, whether we will be able to select for intelligence depends on what the true genetic architecture of the trait is. First, using genetics to predict intelligence across the whole range of the population is one thing—using it to predict the much smaller expected differences between siblings is a totally different proposition, one that would require greater precision than may be obtainable, especially given the influence of developmental variation, which is essentially unpredictable. Second, I described a model in chapter 8 that sees intelligence mainly as a general fitness indicator, reflecting to a large extent the general robustness of brain development and the genomic program that encodes it. If that is true, then intelligence may be determined by general mutational load and the impact of these mutations on brain development, rather than a specific, dedicated set of genes "for intelligence." Selecting for greater intelligence may thus be a matter of choosing embryos with the lowest load of severe mutations likely to affect neural development. Indeed, that would be expected to increase general health as well.

Again, I am not advocating for this, merely laying out the technical parameters. And, in considering it, it is worth remembering the law of unintended consequences. First, any given mutation is likely to have multiple effects on multiple systems—some of these may be unknown or unpredictable and not all of them will necessarily be negative. Second, we are in fact adapted to a certain mutational load—our developmental programs have evolved with such a load in place. We all carry approximately 200 severe mutations—ones that seriously impair production or function of a protein—as well as thousands of less severe genetic variants. And we always have. Every human who ever lived has, just as every animal that has ever lived has had some similar burden. There never has been a human without a certain load of mutations, one that is fully "wild type" across the entire genome. We may have the opportunity to do what natural selection never could—to purge the genome of all such mutations at once, or to reach that point over successive generations. But really we have no idea what the outcome would be—maybe development will proceed perfectly well with all systems working maximally, maybe not. Perhaps we'll all end up super healthy and smart and ridiculously good-looking—and identical.

Genetic information is likely to be used in many areas outside of reproductive decisions too. Perhaps the most obvious is in insurance, where information that predicts people's future health could very well be gleaned from their genomes. This raises serious questions. For example, would just carrying a mutation that statistically increases the risk of developing schizophrenia at a future date be considered a preexisting condition? Would variants that predispose to risky behavior or suicidality be grounds to deny someone life insurance or charge that person higher premiums? Currently, many countries prohibit insurance companies from using such information to deny people coverage (for example, under the Genetic Information Nondiscrimination Act in the United States) but the policies are quite uneven and, of course, could change. Indeed, a bill (H.R. 1313) currently under consideration in the United States (in 2017) would allow employers to demand employees to undergo genetic testing as part of a "wellness" program, or face an increase in their health insurance costs.

It's also not hard to see how genetic information that predicts behavioral traits or cognitive abilities would be of interest to schools, colleges, or employers. IQ and aptitude tests are already widely used—these could conceivably be replaced by genetic predictors. At the moment, such predictors remain hypothetical and they will never be perfect, but they could be developed to the point where they contain some information deemed to be useful in a prospective fashion—say, for streaming children in schools. We could even see the prospect of genetic profiles being used in dating, as depicted in the science fiction film *Gattaca* (along with many of the other scenarios raised here). After all, we already choose mates based on many different traits with genetic underpinnings, and information on such traits is commonly used in selecting sperm or egg donors. Direct-to-consumer genetic profiling is a booming business and is already straying into many of these areas. Science fiction is fast becoming science fact. Buckle up!

A NOTE ON RACE AND GROUP DIFFERENCES

Up to this point, we have been concentrating on the origins of differences between individuals, but have not considered the possibility of average differences between *groups* of individuals, or populations. (With the exception of sex differences, which are a special case, given that there are strong evolutionary reasons for sex differences in behavior and known, conserved mechanisms that institute them.) If psychological traits have a partly genetic basis, so that relatives are more similar in such traits to each other than to random strangers, then it seems reasonable to suppose that such similarity might extend across whole populations who share a common ancestry and cause differences between populations with different ancestries. There are dozens of physical traits—like skin color, facial morphology, or height, for example—that do indeed differ between populations in this way. That this might extend to psychological traits is thus not inconceivable.

However, this is not a given. Systematic differences between groups can sometimes arise just by "genetic drift"—the random divergence between two populations of genetic variants, some of which may affect traits. But that mainly applies to traits that are evolutionarily neutral—where it really doesn't matter much if a trait is high or low. For traits with adaptive value, however, the emergence of systematic differences requires some active force to drive it, some selective advantage to a greater or lower level.

Most of the physical traits that differ between populations have clear adaptive effects—there is a reason that they differ. For example, lighter skin evolved independently a couple of times as humans migrated to more northern latitudes, as an adaptation to lower light conditions. While dark skin is protective in regions with high sunlight, in low light it prevents adequate production of vitamin D. Similarly, persistence into adulthood of expression of the enzyme lactase, which breaks down milk, arose recently (in the past several thousand years) in European populations with the advent of dairy farming. And genetic adaptations to high altitude are seen in some populations, like Tibetans.

However, even if comparable forces did apply for psychological traits (and there is no evidence that they do or have), their genetic architecture makes this kind of directional selection much more difficult. The physical traits mentioned above are driven by changes to one or two genes, with highly specific effects. But we have seen that psychological traits can be affected by genetic variants in hundreds or thousands of genes, which often also affect other traits. That means, first, that any given mutation that increases the level of one trait may have offsetting effects on other traits. This will tend to constrain the possibilities for change. And second, it means that directional selection will face a losing battle against mutation, which will instead constantly generate diversity within groups. There would need to be an extremely strong selective force—similar to the levels of artificial selection that dog breeds were subjected to—in order to drive stable group differences for these kinds of traits.

In addition, for personality traits at least, diversity may actually be *promoted* because there is no single combination of parameters that is optimal in all situations or all environments. Any given profile will lead to more optimal behaviors in some contexts, but less optimal ones in others. For example, in some circumstances cautious people will do better (they may be less likely to get killed, for example). In other situations more daring people may do better (they may be more likely to obtain food or a mating opportunity). Whether one profile outperforms the others in terms of evolutionary fitness depends on how often those different types of situations arise in that particular environment.

But we should remember that the most important thing in each person's environment is other people. Those are the ones we can cooperate or compete with, those are the threats that pose the most danger and the sources of the most relevant opportunities. That means that the optimal profile of behavioral parameters for any individual depends on the profiles of everyone else around that person. Not in a simple way, however; it's not the case that the best solution is to be like everyone else—sometimes quite the opposite. If, for example, most other people are quite reckless, then it may pay to be more cautious. While half of them are dying off because they've put themselves in too much danger, you can hang back and share in the spoils. (It may not be admirable, but natural selection won't care.) If, on the other hand, you're in a population of timid people, you may gain an advantage by being braver, especially in obtaining mating opportunities.

This is classic game theory—the optimal strategy for any individual depends on the strategies employed by others. In evolutionary terms it leads to what is known as frequency-dependent selection. The fitness value of any given phenotype (a behavioral strategy in this case) decreases as the frequency of that phenotype in the population increases beyond a certain point. Any given strategy works better while it's still somewhat rare, which tends to prevent genetic variants that favor any specific behavioral profile from ever getting fixed in the population. Diversity thus arises not just from a fundamental inability to genetically specify the same profile in all individuals but also from the positive actions of natural selection.

So, while a naïve comparison with physical traits suggests that psychological traits might well vary between groups, a more detailed consideration of their genetic architecture reveals just how unusual a scenario would have to exist for this to arise. It is by no means impossible—but it would require strong and consistent environmental differences between groups to create systematic pressures strong enough to drive genetic adaptation for these traits. Which brings us to how such groups are defined and the question of whether the categories typically studied have any real validity.

Most of the discussion in this area centers on the colloquial idea of "races," but exactly how many such categories exist and how they are defined are hard to agree on. Anthropologists in the 1800s identified three main races—Black, White, and Asian—roughly reflecting continental ancestry. But a fourth soon had to be added when it was recognized that Australian aborigines are really very distinct from Africans, despite having similar skin color. And, of course, each of those categories can be subdivided more and more—among Whites, for example, we could recognize Hispanics, Jews, Arabs, etc. In terms of shared ancestry, thousands of such groups can be defined across all areas of the globe. Some will be reasonably discrete, based on a history of isolation and restricted breeding, while others are much more mixed, reflecting more extensive migration and interbreeding.

Modern genetics can reveal much of this history and clearly illustrates the complexities of humanity's global family tree. If you cluster people based on genetic similarity, you can indeed derive several major categories, but you can also just as well go to deeper levels and reveal many, many more. There is no reason to think that any one level should have privileged status—none of these groupings reflects a natural kind, in the way that sex does. You can look for trends at the level of Africans versus non-Africans, for example, but you can also look at the level of ethnic groups like Bantu, Amhara, Yoruba, Celts, Basques, Finns, Japanese, American Indians, Maori, and so on. The decision to stop at any given level of clustering is purely arbitrary, and the larger and more ancient the cluster, the greater diversity there will be *within* that group, both genetically and in terms of the environments to which they have been exposed.

This is an important point when considering claims of racial differences in behavior and the even stronger claim that these are driven by genetic differences. For example, in his 2014 book *A Troublesome Inheritance: Genes, Race and Human History*, journalist Nicholas Wade argues, first, that strong and stable differences in behavioral or cognitive traits exist between five major racial categories and, second, that these are driven by genetic differences, reflecting adaptation to different historical societal structures across continents. As the author admits, such claims are "leaving the world of hard science and entering into a much more speculative arena at the interface of history, economics and human evolution."[2] Quite. It is a complete non sequitur to claim that any cultural differences between populations must be caused by genetic differences. There is in fact no evidence at all that observed or supposed differences in behavioral patterns between populations reflect anything but cultural history.

A more contentious issue is the notion of racial differences in intelligence. The idea that observed differences in cognitive abilities between populations might be driven by genetic differences is an old one, certainly popular with Galton and Davenport, for example. But it achieved notoriety with the publication of the 1994 book *The Bell Curve: Intelligence and Class Structure in American Life*, by psychologist Richard Herrnstein and political scientist Charles Murray. Among other things, they highlighted differences in average scores on IQ tests between various ethnic groups across America, noting a lower average among African-Americans and Hispanics than among Whites or Asians. Since IQ is a heritable trait but can also be affected by environmental factors, they went on to state that: "It seems to us highly likely that both genes and the environment have something to do with racial differences."[3]

This is couched in the most reasonable-sounding terms—simply presenting a "probably a bit of both" scenario as the most likely situation. This seems to put the burden of proof on people who argue that genetic differences will *not* contribute to differences in intelligence across population groups. But is there any evidence for their hypothesis? And is it really likely?

Regarding heritability, twin and family studies only show that much of the variation in IQ *within the studied populations* is due to genetic variation. This says nothing about what might cause differences *between* populations. A trait could be completely heritable within each of two populations yet show a difference between them that is completely environmental. As noted previously, body mass index is highly heritable in both the United States and in France, but the large difference in average body mass index between these countries is caused by environmental factors, not genetic ones.

[2] N. Wade, *A Troublesome Inheritance: Genes, Race and Human History* (New York: Penguin, 2014), 8.
[3] R. J. Herrnstein and C. Murray, *The Bell Curve: Intelligence and Class Structure in American Life* (New York: Free Press, 1994), 311.

In the case of intelligence, we know from trends over time that it is highly sensitive to factors such as general maternal and infant health, nutrition, education, and practices of abstract thinking. Changes to all of these factors have contributed to increases in average IQ scores across many nations over the past century, which have nothing to do with changes in genes. Given the historical and continuing inequities between racial groups in the United States and across the world, it would seem more appropriate to exhaust the possible contributions of these cultural factors before inferring any contribution from genetic differences.

Indeed, behavioral geneticists often rightly criticize sociological studies as being uninterpretable when they don't control for known genetic confounds. For example, the idea that having books in the house causally increases children's IQ is hopelessly confounded by the fact that parents with higher IQ will likely have more books in their house and will also tend to have children with higher IQ, for genetic reasons. The converse is true here. We know that cultural factors affect IQ and we know that they differ very substantially between the groups concerned. The conclusion that differences in IQ test performance reflect, even in part, genetically driven differences in *intellectual potential* across races is thus hopelessly confounded and remains entirely speculative.

But, beyond that, such variation may be inherently *unlikely*. If intelligence is a general fitness indicator rather than a genetically modular trait, this changes the dynamics of possible selection on it. It is not enough to say that greater intelligence might have been selected for in one population—you have to explain why that would *not* have been the case in every population. The selective pressures that led to the emergence of *Homo sapiens* may well have directly favored mutations that led to greater intelligence; that is, selection would have been acting on that trait itself. But once that complex system was in place, the main variation would be in the load of mutations that impair it, which will likely have effects on many traits and impair fitness generally. General fitness should always be selected for, by definition, in any population, meaning intelligence should get a free ride—it will be subject to stabilizing selection, whether or not it is the thing being selected for.

For all these reasons, none of the evidence for genetic effects on psychological traits presented in this book should be taken as supporting the case for a genetic contribution to differences in such traits between populations.

DETERMINISM

I have presented the case in this book for the existence of innate differences in psychological traits, arising from two sources: genetic differences in the program specifying brain development and function, and random variation in how that program plays out in an actual individual. The second source is often overlooked, but its effects mean that many traits are *even more innate* than heritability estimates alone would suggest. In short, we're born different from each other. The slate is most definitively not blank. To many people, this may be the most obvious thing in the world, based on their common experience of other human beings, especially children. To others, however, it may smack of genetic determinism. It may sound like a claim that our genes *determine* our behavior—that we are slaves to them with no real autonomy.

This is not the case at all. The claim is far more modest. It is simply this: that variation in our genes and the way our brains develop cause differences in innate behavioral *predispositions*—variation in our behavioral tendencies and capacities. Those predispositions certainly influence how we behave in any given circumstance but do not by themselves determine it—they just generate a baseline on top of which other processes act. We learn from our experiences, we adapt to our environments, we develop habitual ways of acting that are in part driven by our personality traits, but that are also appropriately context dependent.

Along the same lines, the evidence that parenting does not have a strong influence on our behavioral *traits* should not be taken as implying that parenting does not affect our behavior at all. We may not be molding our children's personalities, but we certainly influence the way they adapt to the world. Our actual behavior at any moment is influenced as much by these characteristic adaptations and by the expectations of family and society—and, indeed, the expectations we build up of ourselves—as by our underlying temperament. Slates don't have to be blank to be written on.

But if I can evade the charge of genetic determinism, I may still appear guilty to some of the related crime of neuroscientific reductionism. In delving into the detailed mechanisms underlying mental functions and what may cause them to vary, it may seem as if I am *reducing* those mental functions to the level of cells and molecules, none of which has a mind or is capable of subjective experience. It may look like such explanations leave no room for real autonomy, for thoughts and ideas and feelings and desires and intentions to have any causal power, for free will to exist at all.

Once again, this is not the case—nothing I have presented in this book is a threat to our general notions of autonomy and free will. The fact that there is a physical mechanism underlying our thoughts, feelings, and decisions does not mean we do not have free will. After all, to expect that thoughts, feelings, and decisions would *not* have any physical substrate is to fall into dualism—the idea that the brain and mind are really fundamentally distinct things, the mind somehow immaterial. This is a fallacy, and one that is hard to climb back out of once you've fallen into it. The mind is not a thing at all—at least, it is not an object. It is a process, or a set of processes—it is, simply put, the brain at work.

Thoughts and feelings and choices are mediated by the physical flux of molecules in the brain, but this does not mean they can be reduced to it. They are emergent phenomena with causal power in and of themselves. Some pattern of neural activity leads to a certain action by virtue of it comprising a thought with some content and meaning for the organism, not merely because the atoms bumping around in a certain way necessarily lead to them bumping around in a new way in a subsequent moment. The precise details of all the atoms don't matter and don't have causal force because most of those details are lost in the processing of information through the neural hierarchy. What matters is the information content inherent in the patterns of neuronal firing that those atoms represent and what that information *means*. When I make a decision it's because my patterns of neural activity at that moment mean something, to me.

We all have predispositions that make us more likely to act in certain ways in certain situations, but that doesn't mean that on any given instance we *have to* act like that. We still have free will, just not in the sense that we can choose to do any old random thing at any moment. I mean, we could, we just usually don't, because we are mostly guided by our habits (which have kept us alive so far) and, when we do make deliberative decisions, it is between a limited set of options that our brain suggests. So, we are not completely free, we are constrained by our psychological nature to a certain extent. But really that's okay—that's what being a self entails. Those constraints are essential for continuity of our *selves* over time. Having free will doesn't mean doing things for no reason, it means doing them for *your* reasons. And it entails moral responsibility in the pragmatic sense that we are judged not just on our actions but also on our reasons for those actions.

This does raise a provocative idea, however—that some of us may have more free will than others. In each one of us our degree of self-control varies in different circumstances, depending on whether we are tired, hungry, distracted, stressed, sleep deprived, intoxicated, infatuated, and so on. And over our lifetimes the impetuosity of youth gives way to the circumspection of adulthood. But the mechanisms that allow us to exercise deliberative control over habitual or reflexive actions also clearly vary in a more trait-like fashion between people. Some people are far more impulsive than others, as we saw in chapter 6. Many suffer from compulsions or obsessions or addictive behavior that they cannot control. And people in the grip of psychosis or mania or depression are clearly not in full control of their actions, which is why we do not hold them legally responsible. You could say that some people are more at the mercy of their biology than others, though that difference itself is a matter of biology.

SELF-HELP

There is a massive self-help industry devoted to the idea that we can change ourselves—our habits, our behaviors, even our personalities. From psychotherapy or cognitive behavioral therapy to mindfulness, brain training, or simply harnessing the power of positive thinking, there are scores of different approaches and an endless supply of books, videos, seminars, and other materials to help you become your best self. These suggest that we can learn the habits of highly effective people, and we too will become highly effective. That we can overcome stress, anxiety, negative thoughts, relationship problems, and low self-esteem, manage our anger, boost our mood, achieve the goals we always hoped for, and generally become a happier person. The slightly paraphrased title of one self-help book promises to show you how to *rewire your brain* to overcome anxiety, boost your confidence, and change your life. Others proclaim that you can "Immediately achieve massive results using powerful (fill in the blank) techniques!"

Lately, what had been an almost exclusively psychological literature has been suffused with supposedly groundbreaking discoveries from neuroscience, which seem to confirm the possibility of change and elucidate the mechanisms by which it can occur. Two areas in particular have caught the public's imagination.

The first is neuroplasticity or brain plasticity—the idea that the structure of the brain is not fixed but quite malleable, with the implication that prewired need not mean hardwired. And this is quite true, to a certain extent. The brain is constantly rewiring itself on a cellular scale—that is how it learns and lays down memories to allow behavioral adaptation based on experience, by forming new synaptic connections between neurons or pruning others away. There is nothing revolutionary about this—it is simply how brains work. It is also true that, after injury for example, the brain can sometimes rewire circuits on a much larger scale, which can aid recovery or compensation for the injury in some cases or lead to additional problems in others.

But the brain is not infinitely malleable, and for good reason—it has to balance the need to change with the need to maintain the physical structure that mediates the coherence and continuity of the self. If it were undergoing wholesale changes all the time we would never be us. While young brains are highly plastic and responsive, these properties diminish drastically beyond a certain stage of maturation—indeed, they are actively held in check by a whole suite of cellular and extracellular changes. The period of plasticity is extremely protracted in humans, reflecting the fact that we have greater cognitive and neural capacity to continue to learn from experience over longer periods of time. But at some point the brain and the individual have to stop becoming and just be.

This limits the amount of change we can expect to achieve. It is certainly possible to change our behaviors—with enough effort you can break a habit or overcome an addiction. And that may be a perfectly laudable and worthwhile goal in many circumstances. But there is little evidence to support the idea that we can really change our personality *traits*, that we could, for example, learn to be biologically less neurotic or more conscientious. You may be able to learn behavioral strategies that allow you to adapt better to the demands of your life, but these are unlikely to change the predispositions themselves.

For children the situation may be different. There may be periods in which intensive behavioral interventions can alter developmental trajectories. For example, a child with autism may be taught to consciously look at people's faces as they are speaking—this may encourage better linguistic and social development than would have tended to occur otherwise. But even here the opportunities to effect long-lasting change are still limited. These kinds of interventions, in either typically or atypically developing children, will always be fighting against both the innate predispositions themselves and their cascading effects on the experiences individuals choose and the environments they select or create, which will tend to reinforce innate traits.

The second idea that is popular these days is known as epigenetics. We came across the word epigenetic in chapter 4, where it was used to refer to the processes of development through which an individual emerges. The modern usage refers to something quite distinct—the molecular mechanisms that cells use to regulate gene expression. In any given cell at any given time, some genes will be active—the proteins they encode will actually be being produced—while others will be silent. This allows muscle cells to make muscle cell proteins and bone cells to make bone cell proteins, and so on. But cells also respond to changes, either internal or external to the cell, by increasing or decreasing the amounts of proteins made from various genes. Epigenetic mechanisms of gene regulation allow these kinds of changes to be locked in place for some period of time, sometimes even through the life of the cell and any cells it produces. That is precisely what happens in development as different cell types differentiate from each other.

The attraction of epigenetics for the self-help industry stems from the idea that it acts as a form of cellular memory, turning genes on or off in response to experience and keeping them that way for long periods of time. The problem comes from thinking that turning genes on or off equates somehow to turning *traits* on or off. If you're talking about something like skin pigmentation, that might apply—I can expose my skin to the sun for a period of time and this will lead to epigenetic changes in the genes controlling pigment production, and I'll get a nice tan that will last for weeks. But for psychological traits, the link between gene action at a molecular level and expression of traits at a behavioral level is far too indirect, nonspecific, and combinatorial for such a relationship to hold. Moreover, if much of the variation in these traits comes from how the brain developed, the idea that you can change them by tweaking some genes in adults becomes far less plausible. So, despite their current cachet, neuroplasticity and epigenetics don't provide any magical means to dramatically alter our psychological traits.

This brings me to a final point, and really it is just my personal opinion. To me, the self-help industry is built on an insidious and even slightly poisonous message. It all sounds very positive—the possibility of change—but really it relies on the idea that you're not good enough as you are, that other people are better than you, but if you buy our products or take our classes or just think positively enough then you can be better too. It plays on some of the least attractive aspects of human psychology, often explicitly using envy as a marketing ploy—of neighbors who've got more money than you, that guy at work who got promoted ahead of you, or that woman who just seems to have the perfect life. And it is often targeted at the more neurotic among us, with claims of overcoming anxiety, worry, stress, low confidence, and low self-esteem, playing on those very personality traits to convince people they need to be changed.

This is not a self-help book—clearly. But perhaps there is something positive in highlighting a different view. There is a power in accepting people the way they are—our friends, partners, workmates, children, siblings, and especially ourselves. People really are born different from each other and those differences persist. We're shy, smart, wild, kind, anxious, impulsive, hardworking, absent-minded, quick-tempered. We literally see the world differently, think differently, and feel things differently. Some of us make our way through the world with ease, and some of us struggle to fit in or get along or keep it together. Denying those differences or constantly telling people they should change is not helpful to anyone. We should recognize the diversity of our human natures, accept it, embrace it, even celebrate it.

BIBLIOGRAPHY

CHAPTER 1

Adams, J. "Genetics of Dog Breeding." *Nat. Educ.* 1 (2008): 144.

Lewens, T. "Human Nature: The Very Idea." *Philos. Technol.* 25 (2012): 459–74.

Machery, E. "A Plea for Human Nature." *Philos. Psychol.* 21 (2008): 321–29.

Pinker, S. *The Blank Slate: The Modern Denial of Human Nature.* New York: Viking, 2002.

Trut, L. "Early Canid Domestication: The Farm-Fox Experiment." *Am. Sci.* 87 (1999): 160–69.

CHAPTER 2

Bouchard, T. J., Jr., and M. McGue. "Genetic and Environmental Influences on Human Psychological Differences." *J. Neurobiol.* 54 (2003): 4–45.

Galton, F. *Inquiries into Human Faculty and Its Development.* 2nd ed. New York: J. M. Dent, 1907.

Harris, J. R. *The Nurture Assumption: Why Children Turn Out the Way They Do.* New York: Free Press, 1998.

Jansen, A. G., S. E. Mous, T. White, D. Posthuma, and T. J. Polderman. "What Twin Studies Tell Us about the Heritability of Brain Development, Morphology, and Function: A Review." *Neuropsychol. Rev.* 25 (2015): 27–46.

Kochunov, P., N. Jahanshad, D. Marcus, A. Winkler, E. Sprooten, T. E. Nichols, S. N. Wright, et al. "Heritability of Fractional Anisotropy in Human White Matter: A Comparison of Human Connectome Project and ENIGMA-DTI Data." *Neuroimage* 111 (2015): 300–311.

Plomin, R., and D. Daniel. "Why Are Children in the Same Family So Different from One Another?" *Int. J. Epidemiol.* 40 (2011): 563–82.

Plomin, R., J. C. DeFries, and G. E. McClearn. *Behavioral Genetics.* 5th ed. New York: Worth, 2008.

Polderman, T.J.C., B. Benyamin, C. A. de Leeuw, P. F. Sullivan, A. van Bochoven, P. M. Visscher, and D. Posthuma. "Meta-analysis of the Heritability of Human Traits Based on Fifty Years of Twin Studies." *Nat. Genet.* 47 (2015): 702–9.

Roshchupkin, G. V., B. A. Gutman, M. W. Vernooij, N. Jahanshad, N. G. Martin, A. Hofman, K. L. McMahon, et al. "Heritability of the Shape of Subcortical Brain Structures in the General Population." *Nat. Commun.* 7 (2016): 13738.

Thompson, P. M., T. Ge, D. C. Glahn, N. Jahanshad, and T. E. Nichols. "Genetics of the Connectome." *Neuroimage* 80 (2013): 475–88.

Turkheimer, E. "Three Laws of Behavior Genetics and What They Mean." *Curr. Dir. Psychol. Sci.* 9 (2000): 160–64.

Visscher, P. M., W. G. Hill, and N. R. Wray. "Heritability in the Genomics Era: Concepts and Misconceptions." *Nat. Rev. Genet.* 9 (2008): 255–66.

Wen, W., A. Thalamuthu, K. A. Mather, W. Zhu, J. Jiang, P. L. de Micheaux, M. J. Wright, et al. "Distinct Genetic Influences on Cortical and Subcortical Brain Structures." *Sci. Rep.* 6 (2016): 32760.

Zhan, L., N. Jahanshad, J. Faskowitz, D. Zhu, G. Prasad, N. G. Martin, G. I. de Zubicaray, K. L. McMahon, M. J. Wright, and P.M. Thompson. "Heritability of Brain Network Topology in 853 Twins and Siblings." *Proc. IEEE Int. Symp. Biomed. Imaging* 2015 (2015): 449–53.

CHAPTER 3

1000 Genomes Project Consortium. "A Global Reference for Human Genetic Variation." *Nature* 526 (2015): 68–74.

Falconer, D. S., and T.F.C. Mackay. *Introduction to Quantitative Genetics.* 4th ed. Harlow, UK: Longmans Green, 1966.

Fisher, R. A. *The Genetical Theory of Natural Selection.* Oxford: Oxford University Press, 1930.

Keller, M. C., and G. Miller. "Resolving the Paradox of Common, Harmful, Heritable Mental Disorders: Which Evolutionary Genetic Models Work Best?" *Behav. Brain Sci.* 29 (2006): 385–404.

Lupski, J. R., J. W. Belmont, E. Boerwinkle, and R. A. Gibbs. "Clan Genomics and the Complex Architecture of Human Disease." *Cell* 147 (2011): 32–43.

Olson, M. V. "Human Genetic Individuality." *Annu. Rev. Genomics Hum. Genet.* 13 (2012): 1–27.

Schrödinger, E. *What Is Life? The Physical Aspect of the Living Cell.* Cambridge, UK: Cambridge University Press, 1944.

Watson, J. *DNA: The Secret of Life.* London: William Heinemann, 2003.

CHAPTER 4

Chang, H. H., M. Hemberg, M. Barahona, D. E. Ingber, and S. Huang. "Transcriptome-wide Noise Controls Lineage Choice in Mammalian Progenitor Cells." *Nature* 453 (2008): 544–47.

Clarke, P. G. "The Limits of Brain Determinacy." *Proc. Biol. Sci.* 279 (2012): 1665–74.

Corey, L. A., J. M. Pellock, M. J. Kjeldsen, and K. O. Nakken. "Importance of Genetic Factors in the Occurrence of Epilepsy Syndrome Type: A Twin Study." *Epilepsy Res.* 97 (2011): 103–11.

Johnson, W., S. W. Gangestad, N. L. Segal, and T. J. Bouchard Jr. "Heritability of Fluctuating Asymmetry in a Human Twin Sample: The Effect of Trait Aggregation." *Am. J. Hum. Biol.* 20 (2008): 651–58.

Kan, K. J., A. Ploeger, M. E. Raijmakers, C. V. Dolan, and H. L. van der Maas. "Nonlinear Epigenetic Variance: Review and Simulations." *Dev. Sci.* 13 (2010): 11–27.

Leamy, J., and C. P. Klingenberg. "The Genetics and Evolution of Fluctuating Asymmetry." *Annu. Rev. Ecol. Evol. Syst.* 36 (2005): 1–21.

Manzini, M. C., and C. A. Walsh. "The Genetics of Cortical Malformations." In *The Genetics of Neurodevelopmental Disorders*, edited by K. J. Mitchell, 129–53. Hoboken, NJ: Wiley Blackwell, 2015.

Marcus, G. *The Birth of the Mind: How a Tiny Number of Genes Creates the Complexities of Human Thought.* New York: Basic Books, 2004.

Mitchell, K. J. "The Genetics of Brain Wiring: From Molecule to Mind." *PLoS Biol.* 5, no. 4 (2007): e113.

Molenaar, P.C.M., D. I. Boomsma, and C. V. Dolan. "A Third Source of Developmental Differences." *Behav. Genet.* 23 (1993): 519–24.

Sanchez, A., S. Choubey, and J. Kondev. "Regulation of Noise in Gene Expression." *Annu. Rev. Biophys.* 42 (2013): 469–91.

Sanes, D., T. Reh, and W. Harris. *Development of the Nervous System.* 3rd ed. London: Academic Press, 2011.

Suárez, R., I. Gobius, and L. J. Richards. "Evolution and Development of Interhemispheric Connections in the Vertebrate Forebrain." *Front. Hum. Neurosci.* 8 (2014): 497.

Vogt, G. "Stochastic Developmental Variation, an Epigenetic Source of Phenotypic Diversity with Far-Reaching Biological Consequences." *J. Biosci.* 40 (2015): 159–204.

Waddington, C. *The Strategy of the Genes.* Bristol, UK: Routledge, 1957.

Wagner, A. *Robustness and Evolvability in Living Systems.* Princeton, NJ: Princeton University Press, 2007.

Wahlsten, D., K. M. Bishop, and H. S. Ozaki. "Recombinant Inbreeding in Mice Reveals Thresholds in Embryonic Corpus Callosum Development." *Genes Brain Behav.* 5 (2006): 170–88.

CHAPTER 5

Avinun, R., and A. Knafo. "Parenting as a Reaction Evoked by Children's Genotype: A Meta-analysis of Children-as-Twins Studies." *Pers. Soc. Psychol. Rev.* 18 (2014): 87–102.

Bates, E., M. H. Johnson, A. Karmiloff-Smith, D. Parisi, and K. Plunkett. *Rethinking Innateness: A Connectionist Perspective on Development.* Cambridge, MA: MIT Press, 1998.

Bick, J., and C. A. Nelson. "Early Adverse Experiences and the Developing Brain." *Neuropsychopharmacology* 41 (2016): 177–96.

Bouchard, T. J., Jr. "Experience Producing Drive Theory: Personality 'Writ Large.'" *Pers. Individ. Differ.* 90 (2016): 302–14.

Briley, D. A., and E. M. Tucker-Drob. "Comparing the Developmental Genetics of Cognition and Personality over the Life Span." *J. Pers.* 85 (2017): 51–64.

Eliot, L. *What's Going On in There? How the Brain and Mind Develop in the First Five Years of Life.* New York: Bantam Books, 1999.

Gottlieb, G. "Experiential Canalization of Behavioral Development: Theory." *Dev. Psychol.* 27 (1991): 4–13.

———. "Probabilistic Epigenesis." *Dev. Sci.* 10, no. 1 (2007): 1–11.

Kuhl, P. K. "Brain Mechanisms in Early Language Acquisition." *Neuron* 67 (2010): 713–27.

LeDoux, J. *Synaptic Self.* New York: Penguin Books, 2002.

Lewis, M. D. "Self-Organizing Individual Differences in Brain Development." *Dev. Rev.* 25 (2005): 252–77.

Scarr, S., and K. McCartney. "How People Make Their Own Environments: A Theory of Genotype-Environment Effects." *Child Dev.* 54 (1983): 424–35.

Teicher, M. H., and J. A. Samson. "Annual Research Review: Enduring Neurobiological Effects of Childhood Abuse and Neglect." *J. Child Psychol. Psychiatry* 57 (2016): 241–66.

CHAPTER 6

Bevilacqua, L., and D. Goldman. "Genetics of Impulsive Behaviour." *Philos. Trans. R. Soc. Lond. B, Biol. Sci.* 368, no. 1615 (2013): 20120380.

Cox, B. R., and J. L. Krichmar. "Neuromodulation as a Robot Controller: A Brain-Inspired Strategy for Controlling Autonomous Robots." *IEEE Robot. Autom. Mag.* 16 (2009): 72–80.

Dalley, J. W., and T. W. Robbins. "Fractionating Impulsivity: Neuropsychiatric Implications." *Nat. Rev. Neurosci.* 18 (2017): 158–71.

Dayan, P. "Twenty-Five Lessons from Computational Neuromodulation." *Neuron* 76 (2012): 240–56.

Dick, D. M., A. Agrawal, M. C. Keller, A. Adkins, F. Aliev, S. Monroe, J. K. Hewitt, et al. "Candidate Gene-Environment Interaction Research: Reflections and Recommendations." *Perspect. Psychol. Sci.* 10 (2015): 37–59.

Doya, K. "Metalearning and Neuromodulation." *Neural Netw.* 15 (2002): 495–506.

Duncan, L. E., and M. C. Keller. "A Critical Review of the First 10 Years of Candidate Gene-by-Environment Interaction Research in Psychiatry." *Am. J. Psychiatry* 168 (2011): 1041–49.

Flint, J., and M. R. Munafò. "Candidate and Non-candidate Genes in Behavior Genetics." *Curr. Opin. Neurobiol.* 23 (2013): 57–61.

Marcinkiewcz, C. A., C. M. Mazzone, G. D'Agostino, L. R. Halladay, J. A. Hardaway, J. F. DiBerto, M. Navarro, et al. "Serotonin Engages an Anxiety and Fear-Promoting Circuit in the Extended Amygdala." *Nature* 537 (2016): 97–101.

Marder, E. "Neuromodulation of Neuronal Circuits: Back to the Future." *Neuron* 76 (2012): 1–11.

Matias, S., E. Lottem, G. P. Dugué, and Z. F. Mainen. "Activity Patterns of Serotonin Neurons Underlying Cognitive Flexibility." *Elife* 6 (2017): pii: e20552.

McAdams, D. P., and J. L. Pals. "A New Big Five: Fundamental Principles for an Integrative Science of Personality." *Am. Psychol.* 61 (2006): 204–17.

McCrae, R. R., P. T. Costa Jr., F. Ostendorf, A. Angleitner, M. Hrebíčková, M. D. Avia, J. Sanz, et al. "Nature over Nurture: Temperament, Personality, and Life Span Development." *J. Pers. Soc. Psychol.* 78 (2000): 173–86.

Nettle, D. *Personality: What Makes You the Way You Are.* Oxford: Oxford University Press, 2009.

Niederkofler, V., T. E. Asher, B. W. Okaty, B. D. Rood, A. Narayan, L. S. Hwa, S. G. Beck, et al. "Identification of Serotonergic Neuronal Modules that Affect Aggressive Behavior." *Cell Rep.* 17 (2016): 1934–49.

Redish, D. *The Mind Within the Brain: How We Make Decisions and How Those Decisions Go Wrong.* New York: Oxford University Press, 2013.

Saucier, G., and S. Srivastava. "What Makes a Good Structural Model of Personality? Evaluating the Big Five and Alternatives." In *APA Handbook of Social and Personality Psychology*, vol. 4, *Personality Processes and Individual Differences*, edited by M. Mikulincer and P. R. Shaver, 283–305. Washington, DC: American Psychological Association, 2015.

Smith, D. J., V. Escott-Price, G. Davies, M. E. Bailey, L. Colodro-Conde, J. Ward, A. Vedernikov, et al. "Genome-wide Analysis of Over 106,000 Individuals Identifies 9 Neuroticism-Associated Loci." *Mol. Psychiatry* 21 (2016): 749–57.

Teissier, A., M. Soiza-Reilly, and P. Gaspar. "Refining the Role of 5-HT in Postnatal Development of Brain Circuits." *Front. Cell Neurosci.* 11 (2017): 139.

Terracciano, A., A. M. Abdel-Khalek, N. Adám, L. Adamovová, C. K. Ahn, H. N. Ahn, B. M. Alansari, et al. "National Character Does Not Reflect Mean Personality Trait Levels in 49 Cultures." *Science* 310 (2005): 96–100.

Yarkoni, T. "Neurobiological Substrates of Personality: A Critical Overview." In *APA Handbook of Social and Personality Psychology*, vol. 4, *Personality Processes and Individual Differences*, edited by M. Mikulincer and P. R. Shaver, 61–83. Washington, DC: American Psychological Association, 2015.

CHAPTER 7

Bargary, G., and K. J. Mitchell. "Synaesthesia and Cortical Connectivity." *Trends Neurosci.* 31, no. 7 (2008): 335–42.

Barnett, K. J., C. Finucane, J. E. Asher, G. Bargary, A. P. Corvin, F. N. Newell, and K. J. Mitchell. "Familial Patterns and the Origins of Individual Differences in Synaesthesia." *Cognition* 106, no. 2 (2008): 871–93.

Cecere, R., G. Rees, and V. Romei. "Individual Differences in Alpha Frequency Drive Crossmodal Illusory Perception." *Curr. Biol.* 25, no. 2 (2015): 231–35.

Clarke, A., and L. K. Tyler. "Understanding What We See: How We Derive Meaning from Vision." *Trends Cogn. Sci.* 19, no. 11 (2015): 677–87.

Deleniv, S. "The Mystery of Tetrachromacy: If 12% of Women Have Four Cone Types in Their Eyes, Why Do So Few of Them Actually See More Colours?" *The Neurosphere* (blog), December 17, 2015. http://theneurosphere.com/2015/12/17/the-mystery-of-tetrachromacy-if-12-of-women-have-four-cone-types-in-their-eyes-why-do-so-few-of-them-actually-see-more-colours/.

Farina, F. R., K. J. Mitchell, and R. J. Roche. "Synaesthesia Lost and Found: Two Cases of Person- and Music-Colour Synaesthesia." *Eur. J. Neurosci.* 45, no. 3 (2017): 472–77.

Galton, F. *Inquiries into Human Faculty and Its Development.* 2nd ed. New York: J. M. Dent, 1907.

Gregory, R. L. *Eye and Brain: The Psychology of Seeing.* Princeton, NJ: Princeton University Press, 1997.

Healy, K., L. McNally, G. D. Ruxton, N. Cooper, and A. L. Jackson. "Metabolic Rate and Body Size Are Linked with Perception of Temporal Information." *Anim. Behav.* 86, no. 4 (2013): 685–96.

Mitchell, K. J. "Curiouser and Curiouser: Genetic Disorders of Cortical Specialization." *Curr. Opin. Genet. Dev.* 21, no. 3 (2011): 271–77.

Newell, F. N., and K. J. Mitchell. "Multisensory Integration and Cross-Modal Learning in Synaesthesia: A Unifying Model." *Neuropsychologia* 88 (2016): 140–50.

Peretz, I. "Neurobiology of Congenital Amusia." *Trends Cogn. Sci.* 20, no. 11 (2016): 857–67.

Sacks, O. *The Man Who Mistook His Wife for a Hat.* London: Picador, 1986.

Samaha, J., and B. R. Postle. "The Speed of Alpha-Band Oscillations Predicts the Temporal Resolution of Visual Perception." *Curr. Biol.* 25, no. 22 (2015): 2985–90.

Schwarzkopf, D. S., C. Song, and G. Rees. "The Surface Area of Human V1 Predicts the Subjective Experience of Object Size." *Nat. Neurosci.* 14, no. 1 (2011): 28–30.

Susilo, T., and B. Duchaine. "Advances in Developmental Prosopagnosia Research." *Curr. Opin. Neurobiol.* 23, no. 3 (2013): 423–29.

Von Uexkull, J. "A Stroll through the Worlds of Animals and Men." In *Instinctive Behavior*, translated and edited by C. Schiller, 5–80. New York: International Universities Press, 1957.

Ward, J. *The Frog Who Croaked Blue: Synesthesia and the Mixing of the Senses*. Hove, UK: Routledge, 2008.

CHAPTER 8

Briley, D. A., and E. M. Tucker-Drob. "Comparing the Developmental Genetics of Cognition and Personality over the Life Span." *J. Pers.* 85, no. 1 (2017): 51–64.

Calvin, W. H. *A Brief History of the Mind*. New York: Oxford University Press, 2004.

Chiang, M. C., M. Barysheva, D. W. Shattuck, A. D. Lee, S. K. Madsen, C. Avedissian, A. D. Klunder, et al. "Genetics of Brain Fiber Architecture and Intellectual Performance." *J. Neurosci.* 29, no. 7 (2009): 2212–24.

Deary, I. "Intelligence." *Annu. Rev. Psychol.* 63 (2012): 453–82.

Ganna, A., G. Genovese, D. P. Howrigan, A. Byrnes, M. Kurki, S. M. Zekavat, C. W. Whelan, et al. "Ultra-Rare Disruptive and Damaging Mutations Influence Educational Attainment in the General Population." *Nat. Neurosci.* 19, no. 12 (2016): 1563–65.

Hill, W. D., G. Davies, S. E. Harris, S. P. Hagenaars, neuroCHARGE Cognitive Working group, D. C. Liewald, L. Penke, C. R. Gale, and I. J. Deary. "Molecular Genetic Aetiology of General Cognitive Function is Enriched in Evolutionarily Conserved Regions." *Transl. Psychiatry* 6, no. 12 (2016): e980.

Kendall, K. M., E. Rees, V. Escott-Price, M. Einon, R. Thomas, J. Hewitt, M. C. O'Donovan, M. J. Owen, J.T.R. Walters, and G. Kirov. "Cognitive Performance among Carriers of Pathogenic Copy Number Variants: Analysis of 152,000 UK Biobank Subjects." *Biol. Psychiatry* 82, no. 2 (2017): 103–10.

Mitchell, K. J. "Genetic Entropy and the Human Intellect." *Trends Genet.* 29, no. 2 (2013): 59–60.

———. "The Genetics of Stupidity." *Wiring the Brain* (blog), July 5, 2012. http://www.wiringthebrain.com/2012/07/genetics-of-stupidity.html.

Okbay, A., J. P. Beauchamp, M. A. Fontana, J. J. Lee, T. H. Pers, C. A. Rietveld, P. Turley, et al. "Genome-wide Association Study Identifies 74 Loci Associated with Educational Attainment." *Nature* 533, no. 7604 (2016): 539–42.

Penke, L., S. M. Maniega, M. E. Bastin, M. C. Valdés Hernández, C. Murray, N. A. Royle, J. M. Starr, J. M. Wardlaw, and I. J. Deary. "Brain White Matter Tract Integrity as a Neural Foundation for General Intelligence." *Mol. Psychiatry* 17, no. 10 (2012): 1026–30.

Pinker, S. "The Cognitive Niche: Coevolution of Intelligence, Sociality, and Language." *Proc. Natl. Acad. Sci. USA* 107, suppl. 2 (2010): 8993–99.

Plomin, R., and I. J. Deary. "Genetics and Intelligence Differences: Five Special Findings." *Mol. Psychiatry* 20 (2015): 98–108.

Ritchie, S. *Intelligence: All that Matters*. London: John Murray Learning, 2015.

Santarnecchi, E., S. Rossi, and A. Rossi. "The Smarter, the Stronger: Intelligence Level Correlates with Brain Resilience to Systematic Insults." *Cortex* 64 (2015): 293–309.

Sniekers, S., S. Stringer, K. Watanabe, P. R. Jansen, J.R.I. Coleman, E. Krapohl, E. Taskesen, et al. "Genome-wide Association Meta-Analysis of 78,308 Individuals

Identifies New Loci and Genes Influencing Human Intelligence." *Nat. Genet.* 49, no. 7 (2017): 1107–12.

Trampush, J. W., M.L.Z. Yang, J. Yu, E. Knowles, G. Davies, D. C. Liewald, J. M. Starr, et al. "GWAS Meta-Analysis Reveals Novel Loci and Genetic Correlates for General Cognitive Function: A Report from the COGENT Consortium." *Mol. Psychiatry* 22, no. 3 (2017): 336–45.

Unz, R. "Race, IQ, and Wealth: What the Facts Tell Us about a Taboo Subject." *American Conservative*, July 18, 2012. http://www.theamericanconservative.com/articles/race -iq-and-wealth/.

Yeo, A., S. G. Ryman, J. Pommy, R. J. Thoma, and R. E. Jung. "General Cognitive Ability and Fluctuating Asymmetry of Brain Surface Area." *Intelligence* 56 (2016): 93–98.

CHAPTER 9

Alanko, K., P. Santtila, N. Harlaar, K. Witting, M. Varjonen, P. Jern, A. Johansson, B. von der Pahlen, and N. K. Sandnabb. "Common Genetic Effects of Gender Atypical Behavior in Childhood and Sexual Orientation in Adulthood: A Study of Finnish Twins." *Arch. Sex. Behav.* 39, no. 1 (2010): 81–92.

Archer, J. "Does Sexual Selection Explain Human Sex Differences in Aggression?" *Behav. Brain Sci.* 32 (2009): 249–311.

Bao, A. M., and D. F. Swaab. "Sexual Differentiation of the Human Brain: Relation to Gender Identity, Sexual Orientation and Neuropsychiatric Disorders." *Front. Neuroendocrinol.* 32, no. 2 (2011): 214–26.

Chekroud, A. M., E. J. Ward, M. D. Rosenberg, and A. J. Holmes. "Patterns in the Human Brain Mosaic Discriminate Males from Females." *Proc. Natl. Acad. Sci. USA* 113, no. 14 (2016): E1968.

Del Giudice, M., T. Booth, and P. Irwing. "The Distance between Mars and Venus: Measuring Global Sex Differences in Personality." *PLoS One* 7, no. 1 (2012): e29265.

Hines, M. *Brain Gender*. New York: Oxford University Press, 2004.

Ingalhalikar, M., A. Smith, D. Parker, T. D. Satterthwaite, M. A. Elliott, K. Ruparel, H. Hakonarson, R. E. Gur, R. C. Gur, and R. Verma. "Sex Differences in the Structural Connectome of the Human Brain." *Proc. Natl. Acad. Sci. USA* 111, no. 2 (2014): 823–28.

Jahanshad, N., and P. M. Thompson. "Multimodal Neuroimaging of Male and Female Brain Structure in Health and Disease across the Life Span." *J. Neurosci. Res.* 95, no. 1–2 (2017): 371–79.

Joel, D., Z. Berman, I. Tavor, N. Wexler, O. Gaber, Y. Stein, N. Shefi, et al. "Sex beyond the Genitalia: The Human Brain Mosaic." *Proc. Natl. Acad. Sci. USA* 112, no. 50 (2015): 15468–73.

Johnson, W., A. Carothers, and I. J. Deary. "Sex Differences in Variability in General Intelligence: A New Look at the Old Question." *Perspect. Psychol. Sci.* 3, no. 6 (2008): 518–31.

Kendler, K. S., L. M. Thornton, S. E. Gilman, and R. C. Kessler. "Sexual Orientation in a US National Sample of Twin and Nontwin Sibling Pairs." *Am. J. Psychiatry* 157, no. 11 (2000): 1843–46.

Knickmeyer, R. C., J. Wang, H. Zhu, X. Geng, S. Woolson, R. M. Hamer, T. Konneker, M. Styner, and J. H. Gilmore. "Impact of Sex and Gonadal Steroids on Neonatal Brain Structure." *Cereb. Cortex* 24, no. 10 (2014): 2721–31.

Långström, N., Q. Rahman, E. Carlström, and P. Lichtenstein. "Genetic and Environmental Effects on Same-Sex Sexual Behavior: A Population Study of Twins in Sweden." *Arch. Sex. Behav.* 39, no. 1 (2010): 75–80.

McEwen, B. S., and T. A. Milner. "Understanding the Broad Influence of Sex Hormones and Sex Differences in the Brain." *J. Neurosci. Res.* 95, no. 1–2 (2017): 24–39.

Puts, D. "Human Sexual Selection." *Curr. Opin. Psychol.* 7 (2015): 28–32.

Ritchie, S. J., S. R. Cox, X. Shen, M. V. Lombardo, L. M. Reus, C. Alloza, M. A. Harris, et al. "Sex Differences in the Adult Human Brain: Evidence from 5,216 UK Biobank Participants." *bioRxiv* (January 22, 2018): 123729.

Ruigrok, A. N., G. Salimi-Khorshidi, M. C. Lai, S. Baron-Cohen, M. V. Lombardo, R. J. Tait, and J. Suckling. "A Meta-Analysis of Sex Differences in Human Brain Structure." *Neurosci. Biobehav. Rev.* 39 (2014): 34–50.

CHAPTER 10

Bao, A. M., and D. F. Swaab. "Sex Differences in the Brain, Behavior, and Neuropsychiatric Disorders." *Neuroscientist* 16, no. 5 (2010): 550–65.

Craddock, N., and M. J. Owen. "The Kraepelinian Dichotomy: Going, Going ... but Still Not Gone." *Br. J. Psychiatry* 196, no. 2 (2010): 92–95.

Deciphering Developmental Disorders Study. "Prevalence and Architecture of De Novo Mutations in Developmental Disorders." *Nature* 542, no. 7642 (2017): 433–38.

Finucane, B., and S. M. Myers. "Genetic Counseling for Autism Spectrum Disorder in an Evolving Theoretical Landscape." *Curr. Genet. Med. Rep.* 4 (2016): 147–53.

Hubbard, L., K. E. Tansey, D. Rai, P. Jones, S. Ripke, K. D. Chambert, J. L. Moran, et al. "Evidence of Common Genetic Overlap between Schizophrenia and Cognition." *Schizophr. Bull.* 42, no. 3 (2016): 832–42.

Jacquemont, S., B. P. Coe, M. Hersch, M. H. Duyzend, N. Krumm, S. Bergmann, J. S. Beckmann, J. A. Rosenfeld, and E. E. Eichler. "A Higher Mutational Burden in Females Supports a 'Female Protective Model' in Neurodevelopmental Disorders." *Am. J. Hum. Genet.* 94, no. 3 (2014): 415–25.

Keller, M. C., and G. Miller. "Resolving the Paradox of Common, Harmful, Heritable Mental Disorders: Which Evolutionary Genetic Models Work Best?" *Behav. Brain Sci.* 29, no. 4 (2006): 385–404.

Mitchell, K. J. (ed.). *The Genetics of Neurodevelopmental Disorders.* Hoboken, NJ: Wiley Blackwell, 2015.

Moreno-de-Luca, A., S. M. Myers, T. D. Challman, D. Moreno-de-Luca, D. W. Evans, and D. H. Ledbetter. "Developmental Brain Dysfunction: Revival and Expansion of Old Concepts Based on New Genetic Evidence." *Lancet Neurol.* 12, no. 4 (2013): 406–14.

Schizophrenia Working Group of the Psychiatric Genomics Consortium. "Biological Insights from 108 Schizophrenia-Associated Genetic Loci." *Nature* 511, no. 7510 (2014): 421–27.

Torres, F., M. Barbosa, and P. Maciel. "Recurrent Copy Number Variations as Risk Factors for Neurodevelopmental Disorders: Critical Overview and Analysis of Clinical Implications." *J. Med. Genet.* 53, no. 2 (2016): 73–90.

Vissers, L. E., C. Gilissen, and J. A. Veltman. "Genetic Studies in Intellectual Disability and Related Disorders." *Nat. Rev. Genet.* 17, no. 1 (2016): 9–18.

CHAPTER 11

Herrnstein, R. J., and C. Murray. *The Bell Curve: Intelligence and Class Structure in American Life*. New York: Free Press, 1994.

Hofstadter, D. *I Am a Strange Loop*. New York: Basic Books, 2007.

Johnson, T., and N. Barton. "Theoretical Models of Selection and Mutation on Quantitative Traits." *Philos. Trans. R. Soc. Lond. B, Biol. Sci.* 360, no. 1459 (2005): 1411–25.

Mitchell, K. J. "Genetic Entropy and the Human Intellect." *Trends Genet.* 29, no. 2 (2013): 59–60.

———. "Top-Down Causation and the Emergence of Agency." *Wiring the Brain* (blog), November 24, 2014. http://www.wiringthebrain.com/2014/11/top-down-causation-and-emergence-of.html.

Murphy, N., G.F.R. Ellis, and T. O'Connor. *Downward Causation and the Neurobiology of Free Will*. Berlin: Springer, 2009.

Olson, M. V. "Human Genetic Individuality." *Annu. Rev. Genomics Hum. Genet.* 13 (2012): 1–27.

Rutherford, A. *A Brief History of Everyone Who Ever Lived: The Stories in Our Genes*. London: Weidenfeld and Nicolson, 2016.

Sperry, R. W. "In Defense of Mentalism and Emergent Interaction." *J. Mind Behav.* 12, no. 2 (1991): 221–46.

Tse, P. U. *The Neural Basis of Free Will: Criterial Causation*. Cambridge, MA: MIT Press, 2013.

Verweij, K. J., J. Yang, J. Lahti, J. Veijola, M. Hintsanen, L. Pulkki-Råback, K. Heinonen, et al. "Maintenance of Genetic Variation in Human Personality: Testing Evolutionary Models by Estimating Heritability Due to Common Causal Variants and Investigating the Effect of Distant Inbreeding." *Evolution* 66, no. 10 (2012): 3238–51.

Wade, N. *A Troublesome Inheritance: Genes, Race and Human History*. New York: Penguin Books, 2015.

Wahlsten, D. "A Contemporary View of Genes and Behaviour: Complex Systems and Interactions." *Adv. Child Dev. Behav.* 44 (2013): 285–306.

Xu-Sheng Zhang, X.-S., and W. G. Hill. "Joint Effects of Pleiotropic Selection and Stabilizing Selection on the Maintenance of Quantitative Genetic Variation at Mutation-Selection Balance." *Genetics* 162 (2002): 459–71.

INDEX

Page numbers *in italics* refer to figures.

nonshared environment. *See* environment, nonshared
noradrenaline, 109, 116
normal distribution. *See* distribution, normal
novelty salience, *110, 115*
NRXN1, 226, 237
nucleus accumbens, *113*
nullisomy, *39*
number form, 144–45, 149
nurturance, 206–207, 209
nurture, 9–10, 12, 14, 84–86; and early brain plasticity, 81–82; and developmental variation, 51, 53–54; and intelligence, 165; nature of, 81–99; and personality traits, 103
nurturing, 186
nutrition, 26, 77, 166–67, 208, 263

obesity, 27, 49
obsessive compulsive disorder, 247
odorant receptor, 134
OFC. *See* cortex, orbitofrontal
olfactory neurons, 134
openness to experience or ideas, 83, 100–101; and educational attainment, 173; sex differences in, 206
opsins, 117, 126, 128, 132, 135, *136*, 136–37
optic nerve, 87, 128
optogenetics, 117–18
ovaries, 188, *189*
oxytocin, 193

pain, 49, 92, 135, 137, 193
pair-bonding, 3, 193
pallidum, 195
parental behaviour or parenting, 2, 28–29, 53, 81, 83–84, *95*, 95–96, 168, 264; in psychiatric disorders, 218, 222; sex differences in, 186, 193
parietal lobe, 70, 175
Parkinson's disease, 176, 247
patterning, 55–57, 61, 64, 69
peas, 31, 37
peers, 52–53, 85, *95*, 96, 198
PER2, 49
perception, 14; as active inference, 126, 128–129, *130*, 130–31; and critical periods, 90; differences in, 133–39; and formation of concepts, 139–154; and learning, 92; in neurodevelopmental disorders, 216, 242; relationship to gene function, 249;

subjective, 125; and synesthesia, 143–154; and the *Umwelt*, 131–33
personality traits, 1, 10, 13–15, 158; classification of, 100–103; and decision-making, 109, *110*, 111–12; developmental influences on, 104, 122–24; and educational attainment, 173; environmental influences on, 53, 83; gene by environment effects on, 121–22; genes and circuits associated with, 105–107; and habits, 264; heritability of, 20, 104; and neuromodulators, 112, 114; in robots, 107–109; selection on, 259; and self-help, 266, 268–69; sex differences in, 187, 199, 205–209; sources of variance of, 103–104
personality disorder, 121, 253
PET (positron emission tomography), 151
PFC. *See* cortex, prefrontal
Phelan-McDermid syndrome, 237
phenotypes, 15, 20, 48–49; prediction of, 252; probabilistic, 54, 70–71, 73, *74*, 76–77, 222, 244; relationship to genotype, 27, 250, 252; sex differences in variance of, 208, 211; of synesthesia, 148–50
phenylketonuria, 246
phenylthiocarbamide, 134
phoneme, 91, 149
photoreceptor, 117, 126, *127*, 135
physiological states, 92, 111
Pinker, Steven, 1, 28
pioneer axons, *63*, 71, *72*
PISA. *See* Programme for International Student Assessment
plasticity: brain or synaptic, 59, 80–82, 86, *88*, 89–91, 93, *94*, 111, 118, 267, 269; and emergence of phenotypes, 198, *251*; gating by neuromodulators, 111, 118, *119*, 120, 174; in neurodevelopmental disorders, 237, 242, *243*, 246; reduction with maturation, 99; sex differences in, 193, 199
Plato, 1, 171
play, rough-and-tumble, 204–205
positive thinking, 266
possession, demonic, 218
post-traumatic stress disorder, 83
predispositions, behavioral or innate, 13, 29, 81–82, 95; and determinism, 264–65; and self-help, 268
primates, 156; sex differences in, 184, 186, 193, 198
probabilistic relationship between genotypes and phenotypes, 70, *72*, 73, 75, 77, 148, 202, 222, 252